Primary **Maths**
for **Scotland**

the education publisher
for **Scotland**

Early Level Maths

Assessment Pack

Authors: Sheena Dunlop and Lesley Ferguson
Assessment Consultant: Carol Lyon
Series Editor: Craig Lowther

© 2020 Leckie

001/10102020

10 9 8 7 6 5

ISBN 9780008392468

Published by
Leckie
An imprint of HarperCollinsPublishers
Westerhill Road, Bishopbriggs, Glasgow, G64 2QT
T: 0844 576 8126   F: 0844 576 8131
leckiescotland@harpercollins.co.uk
www.leckiescotland.co.uk

HarperCollins Publishers
Macken House, 39/40 Mayor Street Upper, Dublin 1,
D01 C9W8, Ireland

Publisher: Fiona McGlade
Project manager: Rachel Allegro
Special thanks to
Copy editor: Mitch Fitton
Layout and illustration: Jouve
Proofreader: Dylan Hamilton

A CIP Catalogue record for this book is available from the British Library

Printed and Bound in the UK using 100% Renewable Electricity at CPI Group (UK) Ltd

MIX
Paper | Supporting responsible forestry
FSC™ C007454

This book is produced from independently certified FSC™ paper to ensure responsible forest management.

For more information visit: www.harpercollins.co.uk/green

Contents

Find all the downloadable resources here: collins.co.uk/primarymathsforscotland

Introduction

This Assessment Pack includes high quality assessments for Numeracy and Mathematics. It can be used to support ongoing professional judgements of children's progress and next steps within the Scottish Curriculum for Excellence (CfE) Early Level. Assessment Packs for First and Second Levels are also available, to ensure consistency from preschool to P7.

As stated in *Building the Curriculum 5: A framework for assessment*, we at Leckie understand that ongoing assessment and pupil feedback is vital to "…support learning and promote learner engagement resulting in greater breadth and depth in learning, including a greater focus on the secure development of knowledge, understanding and skills." Scottish Government (2011).

In this introduction, you will find:

1. Guidance for assessment

2. Introduction to the assessments supplied in this pack

3. Support for record keeping

4. Tables showing coverage of the Experiences and Outcomes and benchmarks

Guidance for assessment

Ongoing formative assessment is an essential part of the learning process. Black and Wiliam (2009) listed five key principles of learning:

1. Clarifying and sharing learning intentions and success criteria;

2. Engineering effective classroom discussions and other learning tasks that elicit evidence of student understanding;

3. Providing feedback that moves learners forward;

4. Activating students as instructional resources for one another; and

5. Activating students as the owners of their own learning.

The Leckie Primary Maths for Scotland Textbooks and Teacher Guides were designed and written with these principles at the forefront, to support teachers to plan and deliver excellent mathematical learning. Both resources create opportunities for: diagnostic questions that assess prior learning; observation of and feedback on day-to-day learning (including outdoor learning); leading and listening to learning conversations.

The Leckie Primary Maths for Scotland Assessment Packs now add to this by providing rich, diagnostic tasks and activities to help the educator track and monitor progress and achievement. The materials contained within the packs are both summative and formative in nature. Not only do they provide important information about the child's current knowledge and skills (summative information) they also help diagnose misconceptions and identify next steps for both the educator and child (formative information).

The Leckie Teacher Guides contain advice and support that will help the educator address common misconceptions that children experience as they learn and develop mathematical skills, thus helping to ensure smooth and continuous progress.

Throughout the learning, teaching and assessment process, feedback should be used to ensure the pupil can answer these questions:

• Where am I going?

• How am I going there?

• Where to next?

Hattie, J. (2012)

Introduction

Primary Maths for Scotland Early Level Assessment Pack

The Primary Maths for Scotland Early Level Assessment Pack contains two types of assessment:

1. Yearly progress checks

2. End of level assessments

Yearly progress checks

These checks provide full coverage of the CfE Early Level Experiences and Outcomes and assess procedural fluency, recall and understanding of key numeracy and mathematical concepts.

When to use the yearly progress checks

There are two yearly progress checks included in this pack. Yearly progress check 1 is designed to be used at the end of preschool. Yearly progress check 2 is designed to be used at the end of Early Level which, for almost all children, will be at the end of P1.

How to use the yearly progress checks

Yearly progress check 1 can be completed in a one-to-one setting or in a small group. There is no writing required of the child, but an adult must ask the questions and record the responses. Yearly progress check 2 can be completed in groups or by the whole class as it is mostly a write-on assessment.

The questions have been arranged in the order of the Experiences and Outcomes for ease of tracking. The educator will use professional judgment in deciding how the assessment is administered, for example in one session or over a period of several days. For both assessments, we have provided suggestions for grouping questions together.

The yearly progress checks are a blend of questions where the child will need to verbalise their answer and explain their thinking, and questions where the child will show you their understanding using manipulatives and written answers. There are photocopiable resources online for ease of use.

The yearly progress checks include a mix of bare number problems, questions using pictures and diagrams and contextualised problems. Throughout the assessment, the pupil is encouraged to explain their thinking to help you to understand the strategy they are using. This is important as the same question can be solved in many, increasingly sophisticated ways.

At times, the pupil will become stuck on a question; this is to be expected. It may be appropriate for you to read or restate the question, perhaps in a context that the pupil is more familiar with. Be mindful, however, not to make the question easier or to give away a strategy that the pupil can use. Wherever a pupil struggles, this is evidence of next steps in learning.

The answers the pupil provides will be an indication of the mathematical and numerical facts that they have committed to memory and the strategies and skills they are using to solve problems. Misconceptions will also be highlighted by the responses given. The Primary Maths for Scotland Teacher Guides list possible misconceptions by topic and provide help and suggested tasks for the pupil to undertake. By comparing the data collected from a group or class it will also be possible to check how well a particular topic was learned and whether the issue lies with the individual or the group.

Using the marking guidance

The marking guidance provided for the yearly progress checks includes a reference to the appropriate Experience and Outcome(s) linked to each question. The 'Notes' section explains what to expect of a child who is 'on track' within the level and provides advice on 'further learning required' for the child who has struggled or could not answer the question. Where further learning is required, the Leckie Teacher Guide describes common misconceptions and offers advice on appropriate learning activities to help address these.

Introduction

Question:

8. **a** Give the child 18 aliens (toys or photocopy and cut out the aliens from Resource 13 twice). **Count the aliens. How many aliens are there? Now write the total in the box.**

   b Now move the aliens into a circle. **How many aliens are there?**

Marking guidance:

Topic – benchmarks / Es & Os	Question	Notes
Number – order and place value MNU 0-02a	8	**On track** • The child recognises that they have to count out the exact amount asked for. • They work systematically, counting each piece of treasure as it is collected. • They count correctly matching one number name to each piece of treasure retrieved from the chest. • They recognise when the correct amount has been reached and so stop counting at five. **Further learning required** • The child is unsure of the forward counting sequence and so forgets how many pieces of treasure they were asked to get. • They do not coordinate each number name with one object only and instead randomly grab a handful of treasure.

End of level assessments

These assessments are contextualised, holistic assessments that blend together topics from across the Mathematics and Numeracy Curriculum. There are several questions per assessment that can be used individually or together to provide a broad picture of the pupil's mastery of the national standards set out in the benchmarks.

The questions will be in new and unfamiliar contexts to the pupil. This will increase the challenge for them as they have to both understand the context and work out what strategy, skills and knowledge to use. As with the yearly progress checks, the pupil is encouraged to show their thinking to help you understand the strategies they are using and identify any misconceptions.

When to use the end of level assessments

End of level assessments should be used when you judge that a child or group of children has mastered the Early Level Mathematics Curriculum, or to ascertain whether they have achieved mastery in instances where you are unsure. This will generally be at the end of P1 if the child is meeting age-related expectations.

How to use the end of level assessments

The end of level assessments are a blend of activities and questions linked to a context. It is advisable that the children are introduced to the context before they are assessed on their Numeracy and Mathematical skills and knowledge. This can be done through storytelling, playing games, discussion and linking the assessment activity to previous learning.

Assessments can be completed in groups or as individuals. When children complete the assessment tasks in groups, they will discuss their ideas and allow the teacher to observe the skills and knowledge they have mastered. Throughout the assessment children can use concrete materials, drawings, diagrams etc. if necessary.

Introduction

Using the marking guidance

The marking guidance for the end of level assessments includes the Experiences and Outcomes for each question part, the question, a description of the answer that tells you the pupil is 'on track' and a 'review' section that describes what you can expect a child to do if they have not mastered the concept assessed. The Leckie Teacher Guide has further guidance on how to support pupils who were unable to engage in the assessment tasks, to help move their learning forwards.

| **Task 1** Happy Birthday, Teddy | MNU 0-10a
I am aware of how routines and events in my world link with times and seasons, and have explored ways to record and display these using clocks, calendars and other methods.

MNU 0-20a
I can collect objects, and ask questions to gather information, organising and displaying my findings in different ways. | <ul><li>Links daily routines and personal events to time sequences.</li><li>Names the days of the week in sequence, knows the months of the year and talks about features of the four seasons in relevant contexts.</li><li>Asks simple questions to collect data for a specific purpose.</li><li>Applies counting skills to ask and answer questions and makes relevant choices and decisions based on the data.</li><li>Contributes to concrete or pictorial displays where one object or drawing represents one data value, using digital technologies as appropriate.</li><li>Interprets simple graphs, charts and signs and demonstrates how they support planning, choices and decision-making.</li></ul> |

Record keeping

Yearly progress checks record keeping

There are record sheets for you to use to track the progress of the pupils against each question. We suggest that you 'mark' each child as they complete their Early Level yearly progress check. We suggest the ABC coding described below but this can easily be replaced with green/amber/red to provide a quick visual check:

A. Chose an appropriate strategy/method and used it correctly (green)

B. Chose an appropriate strategy/method but used it incorrectly or made an error in calculation (amber)

C. Chose an inappropriate strategy/method or did not attempt the question (red)

End of level assessment record keeping

We have provided record sheets for you to use as the children complete the end of level assessments and suggest using the coding: 'O. On track' and 'R. Review'.

A colour coding system can be used to provide a quick visual check, e.g. green for 'on track' and red for 'review'.

It is important to note that a child does not need to successfully complete every task contained within the Leckie Primary Maths for Scotland Assessment Pack in order to achieve CfE Early Level. This decision will be based on the teacher's professional judgement as they observe their pupils engaged in their day-to-day learning. Nevertheless, we hope the record sheets included will provide a simple reference to where children have mastery of key Numeracy and Mathematics concepts and where they require further support.

References

Black P. and Wiliam D. (2009) Developing the Theory of formative Assessment. Educational Assessment, Evaluation and Accountability, 21(1), 5-31

Hattie, J.A.C. (2012) Visible Learning for Teachers. London. Routledge

Scottish Government (2011) Building the Curriculum 5 – A framework for Assessment available online https://www.education.gov.scot/Documents/btc5-framework.pdf accessed on 06/05/2020

Introduction

Tables showing coverage of the CfE Numeracy and Mathematics Experiences and Outcomes and benchmarks

Curriculum organisers	Experiences and Outcomes	Benchmarks to support practitioners' professional judgement of achievement of a level	Coverage in Early Level Assessment Pack
Estimation and rounding	I am developing a sense of size and amount by observing, exploring, using and communicating with others about things in the world around me. MNU 0-01a	• Recognises the number of objects in a group, without counting (subitising) and uses this information to estimate the number of objects in other groups. • Checks estimates by counting. • Demonstrates skills of estimation in the contexts of number and measure using relevant vocabulary, including less than, longer than, more than and the same.	YPC 1 YPC 2 EOL 1, 3, 5, 6
Number and number processes	I have explored numbers, understanding that they represent quantities, and I can use them to count, create sequences and describe order. MNU 0-02a I use practical materials and can 'count on and back' to help me understand addition and subtraction, recording my ideas and solutions in different ways. MNU 0-03a	• Explains that zero means there is none of a particular quantity and is represented by the numeral 0. • Recalls the number sequence forwards within the range 0 – 30, from any given number. • Recalls the number sequence backwards from 20. • Identifies and recognises numbers from 0 to 20. • Orders all numbers forwards and backwards within the range 0 – 20. • Identifies the number before, the number after and missing numbers in a sequence within 20. • Uses one-to-one correspondence to count a given number of objects to 20. • Identifies 'how many?' in regular dot patterns, for example, arrays, five frames, ten frames, dice and irregular dot patterns, without having to count (subitising). • Groups items recognising that the appearance of the group has no effect on the overall total (conservation of number). • Uses ordinal numbers in real life contexts, for example, 'I am third in the line'. • Uses the language of before, after and in-between. • Counts on and back in ones to add and subtract.	YPC 1 YPC 2 EOL 2, 3, 4, 5, 6 YPC 1 YPC 2 EOL 2, 3, 4, 5, 6

Introduction

Curriculum organisers	Experiences and Outcomes	Benchmarks to support practitioners' professional judgement of achievement of a level	Coverage in Early Level Assessment Pack
		• Doubles numbers to a total of 10 mentally. • When counting objects, understands that the number name of the last object counted is the name given to the total number of objects in the group. • Partitions quantities to 10 into two or more parts and recognises that this does not affect the total. • Adds and subtracts mentally to 10. • Uses appropriately the mathematical symbols +, − and = • Solves simple missing number problems.	
Fractions, decimal fractions and percentages	I can share out a group of items by making smaller groups and can split a whole object into smaller parts. MNU 0-07a	• Splits a whole into smaller parts and explains that equal parts are the same size. • Uses appropriate vocabulary to describe halves. • Shares out a group of items equally into smaller groups.	YPC 1 YPC 2 EOL 1, 2, 6
Money	I am developing my awareness of how money is used and can recognise and use a range of coins. MNU 0-09a	• Identifies all coins to £2. • Applies addition and subtraction skills and uses 1p, 2p, 5p and 10p coins to pay the exact value for items to 10p.	YPC 1 YPC 2 EOL 1, 2, 3
Time	I am aware of how routines and events in my world link with times and seasons, and have explored ways to record and display these using clocks, calendars and other methods. MNU 0-10a	• Links daily routines and personal events to time sequences. • Names the days of the week in sequence, knows the months of the year and talks about features of the four seasons in relevant contexts. • Recognises, talks about and where appropriate, engages with everyday devices used to measure or display time, including clocks, calendars, sand timers and visual timetables. • Reads analogue and digital o'clock times (12 hour only) and represents this on a digital display or clock face. • Uses appropriate language when discussing time, including before, after, o'clock, hour hand and minute hand.	YPC 1 YPC 2 EOL 2, 4, 6

Introduction

Curriculum organisers	Experiences and Outcomes	Benchmarks to support practitioners' professional judgement of achievement of a level	Coverage in Early Level Assessment Pack
Measurement	I have experimented with everyday items as units of measure to investigate and compare sizes and amounts in my environment, sharing my findings with others. MNU 0-11a	• Shares relevant experiences in which measurements of lengths, heights, mass and capacities are used, for example, in baking. • Describes common objects using appropriate measurement language, including tall, heavy and empty. • Compares and describes lengths, heights, mass and capacities using everyday language, including longer, shorter, taller, heavier, lighter, more and less. • Estimates, then measures, the length, height, mass and capacity of familiar objects using a range of appropriate non-standard units.	YPC 1 YPC 2 EOL 1, 3, 4, 5
Patterns and relationships	I have spotted and explored patterns in my own and the wider environment and can copy and continue these and create my own patterns. MTH 0-13a	• Copies, continues and creates simple patterns involving objects, shapes and numbers. • Explores, recognises and continues simple number patterns. • Finds missing numbers on a number line within the range 0 – 20.	YPC 1 YPC 2 EOL 1, 3, 4, 5, 6
Properties of 2D shapes and 3D objects	I enjoy investigating objects and shapes and can sort, describe and be creative with them. MTH 0-16a	• Recognises, describes and sorts common 2D shapes and 3D objects according to various criteria, for example, straight, round, flat and curved.	YPC 1 YPC 2 EOL 1, 3, 5
Angle, symmetry and transformation	In movement, games, and using technology I can use simple directions and describe positions. MTH 0-17a	• Understands and correctly uses the language of position and direction, including in front, behind, above, below, left, right, forwards and backwards, to solve simple problems in movement games.	YPC 1 YPC 2 EOL 1, 3, 4
	I have had fun creating a range of symmetrical pictures and patterns using a range of media. MTH 0-19a	• Identifies, describes and creates symmetrical pictures with one line of symmetry.	YPC 1 YPC 2 EOL 1, 4

Introduction

Curriculum organisers	Experiences and Outcomes	Benchmarks to support practitioners' professional judgement of achievement of a level	Coverage in Early Level Assessment Pack
Data and analysis	I can collect objects and ask questions to gather information, organising and displaying my findings in different ways. MNU 0-20a I can match objects, and sort using my own and others' criteria, sharing my ideas with others. MNU 0-20b I can use the signs and charts around me for information, helping me plan and make choices and decisions in my daily life. MNU 0-20c	• Asks simple questions to collect data for a specific purpose. • Collects and organises objects for a specific purpose. • Applies counting skills to ask and answer questions and makes relevant choices and decisions based on the data. • Contributes to concrete or pictorial displays where one object or drawing represents one data value, using digital technologies as appropriate. • Uses knowledge of colour, shape, size and other properties to match and sort items in a variety of different ways. • Interprets simple graphs, charts and signs and demonstrates how they support planning, choices and decision making.	YPC 1 YPC 2 EOL 1, 2, 3 YPC 1 YPC 2 EOL 1 EOL 3, 4

Yearly progress check 1: end of preschool

Teacher instructions

Context

Set up the area with a pirate theme and explain to the children that they are going to help Pirate Penny with some maths problems. Each activity can be done individually or a few activities can be grouped together. Suggested groupings of questions are shown below. All activities should be done one-to-one and observations noted. Resources are provided although toy resources could be used in their place.

Suggested grouping of questions

1–2 • 3–5 • 6 • 7–11 • 12 • 13–14 • 15–16 • 17–18 • 19–20 • 21–23 • 24 • 25 • 26 • 27 • 28–29 • 30–31

Resources

Resources provided online:

Digit cards 0–10 • Dot cards 1–5 • Number track • Coins • Pirate sequence cards • Treasure chest • Treasure map • Pirate flags • Selection of pirates

Resources needed and available in the setting:

Paper and (coloured) pencils • Pieces of treasure (e.g. coins, gems, pieces of silver foil scrunched up to look like silver nuggets) • Bags for coins • Toy pirates or pictures of pirates • Coloured beads and a string • Washing line and pegs • Treasure feely bag • 2D shapes and 3D objects • Real coins – 1p, 2p, 5p, 10p, 20p, 50p, £1, £2 • 3 necklaces (of different lengths) • 3 jars (of different capacities) • 3 bags of coins (of different weights) • Balance scales • Funnels and rice • Box / container / treasure chest

Questions

1. Say to the child **I would like you to start at zero and keep counting up to twenty.** Note if they say **all** the number names in the correct order.

2. Say to the child **I would like you to start at ten and count down to zero.** Note if they say **all** the number names in order.

3. Have a stack of digit cards from 0 to 10 ready. Turn over one card at a time, in random order, and ask **What is this number? ...And this one, ...And this next number, etc**. Note if the child can recognise each numeral.

4. Lay out the digit cards from 1 to 5 randomly on the desk. Say **Put the numbers in order from smallest to biggest.** Once complete take in the cards and lay out the digit cards from 0 to 10 in a random order and say **Put these numbers in order from smallest to biggest.** Note if the order is correct.

5. Have a stack of digit cards from 0 to 10 ready. Choose a card at random and ask the child **What number is this? What number comes before/after?** Repeat with different numerals. Note if the child just knows the number before/after or has to 'drop back'.

6. Give the child a piece of blank paper. Ask the child to **Write each number as I say it.** Starting at zero say the number names, in order, and give them time to write each digit. Note their ability to write the correct numeral from you saying it aloud. Also note their accuracy in writing the numeral.

7. Lay out five pieces of treasure. Ask the child to **Count how many pieces of treasure there are**. Note if the child counts each object once, correctly matching one number name to each object. Note if they touch each object or do this mentally. If the child does this mentally ask them to repeat the activity with nine pieces of treasure. It is acceptable for the child to move the objects to help them count accurately.

8. Lay out a box of treasure. Ask the child to **Bring me five pieces of treasure**. Note if the child counts the objects out one at a time, saying the number names in the correct order.

9. Lay out six pieces of treasure, randomly. Ask the child to **Count how many pieces of treasure there are**. Once the child has counted the objects ask **So how many pieces of treasure are there?** Note if the child can state the answer straightaway or if they need to re-count the objects to tell you the answer.

10. Using the six pieces of treasure just counted, move the objects into a straight line. Ask **How many pieces of treasure are there?** Note if the child knows that changing the position of the objects does not change the total, and says the amount straight away, or if they have to re-count.

11. Give the child a bag of gold coins. Say **Take out five coins**. Note what they do and then put the coins back in the bag. **Now take out zero coins**. Again, take a note of what the child does/says. **Now take out eight coins.** Note if the child can explain that zero means none. Also note if the child is using one-to-one correspondence.

12. Have dot cards from one to five ready (you could use Resource 2). The dot cards required are the dice patterns for the numbers 1–5. Say **I am going to show you some cards with dot patterns on them. Tell me how many dots you can see as soon as you know.** Show the child each card for 2 seconds only. Encourage them to say the number as soon as they know. Note if they can identify the amount without counting (i.e. they can subitise).

13. Show the child two piles of coins. One with three coins and one with six coins. Ask **Which pile has more?** Repeat with different but similar amounts and ask **Which pile has less?** Ask the child to justify their answers. Note their comments and use of mathematical language. Using two piles of coins, first ask the child to **Estimate how many coins are in each pile**. Then ask the child to **Count each amount**. Note the accuracy of the child's estimate. Also note if the child can find the corresponding numeral on a number track (see Resource 6).

14. Using coins ask the child to **Make two collections of equal size by matching one-to-one.** Note if the child can match each collection one-to-one ensuring they have equal amounts.

15. Lay out two collections of coins – one collection with two coins and the other with three coins. Ask the child to **Count each collection separately** then ask **How many altogether?** Note how the child finds the answer, do they count all or count on? Ask them **Explain what you did.**

16. Lay out four coins. Ask the child to **Count the coins and say how many.** Ask the child to **Hide two coins in the treasure chest** (see Resource 8). Ask the child **How many coins are left?** Ask them **Explain what you did.**

17. Ask the child to **Share six coins fairly between two pirates**. Note if the child shares the coins equally, giving three to each pirate, and ask them to justify their answer.

18. Use six coins. Ask the child **Show me all the ways you can partition the coins.** Note if the child partitions the collection into two or more groups and whether they know that partitioning does not affect the total.

19. Tell the child **The pirates find a necklace**. Give the children beads and a string and ask them to **Make a pattern**. **Tell me about the pattern you have made.** Note if they can create a pattern and how complex it is.

20. Set up a washing line with the numbers 3 to 7 (you could use Resource 1) in order. Ask the child to close their eyes while you turn one number round and then ask the child **Which number is missing?** Note if the child can tell you the missing number immediately or if they have to drop back. Note also if the child can use the language before/after and in-between to talk about the number pattern.

21. Lay out a line of five pirates. Ask the child **Point to the third pirate; the first pirate; last pirate.**

22. Hold up one pirate and then say **One more pirate comes along. How many now?** Repeat for two, three, four, five pirates. Now say **One pirate goes away, how many now?** Repeat until there are no pirates left.

23. Place two pirates on the table. Say **Show me double the number of pirates.** Repeat for three, four and five pirates.

24. Present the child with a treasure feely bag containing a circle, triangle, square and rectangle. Say **Take out a shape. Do you know what this shape is called? What can you tell me about your shape?** Note if the child can name the shape and use words like straight, round, flat, curved, etc., to describe it. Repeat the activity for objects such as cylinder, sphere, cone, cube and cuboid.

25. Give the child a treasure chest with real money in it. Discuss with the child what money is and what it is used for. Ask the child **Take out the coins. What can you tell me about them?** Note if the child understands what money is and how it works. Also note if they can identify any specific coins and any money-related language they may use.

26. Show the child pictures of a pirate getting ready for bed (see Resource 7). Ask the child to **Order the pictures.** Discuss what they do when they get ready for bed. Note if the child can use language such as first, next, then, and can link this action to a time of the day.

27. Give the child a large treasure chest with three different lengths of necklace, three coloured jars of different capacity and three different weighted bags of coins. Ask the child the following questions – **Give me the longest necklace, the shortest necklace, the lightest bag of coins, the heaviest bag of coins, the jar that will hold the most and the jar that will hold the least.** Note the child's ability to understand the language and complete the task. Do they measure or go by feel/look? Have measuring equipment available if the child requests it.

28. Show the child the treasure map (see Resource 9). Ask the following questions – **What is beside the tree? What is below the pond? What is above the hill?** Ask the child to **Put a coin on the map. Now tell me about where you have put it.** Note the child's ability to understand and use positional language including in front, behind, above, below, left, right, forwards and backwards.

29. Show the child the pictures of the pirate flags (see Resource 10). Ask the child **Which flag do you like? What do you like about it?** If they identify the symmetrical flag, ask **What do you notice about it?** If appropriate, introduce the word 'symmetrical'. Note their ability to recognise symmetry.

30. Lay out six gold and seven silver coins (see Resource 12) randomly beside two pirates. Tell the child **One pirate owns all the gold coins. The other pirate owns all the silver coins.** Explain to the child **We want to find out which pirate has more coins.** Note how the child starts sorting the coins and gathering information.

31. Give the child five pirates. The pirates have certain things in common and some differences. Ask the child to **Sort the pirates by choosing a common feature and explain how they did it.** Note if they can sort the pirates and give a reason for the sorting.

Topic – benchmarks / Es & Os	Question	Notes
Number – order and place value MNU 0-02a	1	**On track** • The child says each number in the correct order and completes the forward counting sequence correctly. **Further learning required** • The child omits or repeats a number or numbers. • They mix up the number sequence. • They stop counting and cannot continue.
Number – order and place value MNU 0-02a	2	**On track** • The child says each number in the correct order and completes the backward counting sequence correctly. **Further learning required** • The child omits or repeats a number or numbers. • They mix up the number sequence. • They stop counting and cannot continue.
Number – order and place value MNU 0-02a	3	**On track** • The child can immediately recognise the numeral and say the corresponding number name. **Further learning required** • The child does not recognise the numeral and cannot give its number name. • The child gives the wrong number name for the numeral.
Number – order and place value MNU 0-02a	4	**On track** • The child understands that the numbers symbolise quantities and can use this knowledge to confidently order them. **Further learning required** • They need to use the 0–10 counting sequence to order the numbers. • They require a visual support, for example, a number track, so that they can 'match' each digit card with its position in the number sequence.

Topic – benchmarks / Es & Os	Question	Notes
Number – order and place value MNU 0-02a	5	**On track** • The child can identify the number before or after immediately, without the need to 'drop back'. **Further learning required** • The child can confidently give the number after but not the number before. • They have to 'drop back' and say the number sequence up to the number given to enable them to identify the number before and the number after. • They require a visual reference, for example, a number track.
Number – order and place value MNU 0-02a	6	**On track** • The child can picture the symbol which corresponds to each number name said and create recognisable versions of them on paper. **Further learning required** • The child cannot write recognisable symbols for any or some of the number names said. • They may interpret the request to 'write a number' as a request to draw a number of items. For example, when asked to 'write 3' the child draws three things.
Number – order and place value MNU 0-02a	7	**On track** • The child works systematically, counting each piece of treasure once. • They count correctly, matching one number name to each piece of treasure. • They recognise that the number name of the last piece of treasure counted also gives the total for the group. **Further learning required** • The child mixes up the forward counting sequence. • They are unable to coordinate each number name with one object only, counting objects more than once or missing objects. • They do not recognise that the last number said tells them how many there are altogether.

Topic – benchmarks / Es & Os	Question	Notes
Number – order and place value MNU 0-02a	8	**On track** • The child recognises that they have to count out the exact amount asked for. • They work systematically, counting each piece of treasure as it is collected. • They count correctly matching one number name to each piece of treasure retrieved from the chest. • They recognise when the correct amount has been reached and so stop counting at five. **Further learning required** • The child is unsure of the forward counting sequence and so forgets how many pieces of treasure they were asked to get. • They do not coordinate each number name with one object only and instead randomly grab a handful of treasure.
Number – order and place value MNU 0-02a	9	**On track** • The child works systematically, counting each piece of treasure once. • They count correctly matching one number name to each piece of treasure. • They recognise that the number name of the last piece of treasure counted also gives the total for the group. **Further learning required** • The child mixes up the forward counting sequence. • They are unable to coordinate each number name with one object only, counting objects more than once or missing objects. • They need to re-count when asked, 'So how many are there?'
Number – order and place value MNU 0-02a	10	**On track** • The child understands that moving the objects doesn't change the total and immediately gives the correct answer. **Further learning required** • They do not trust that if no objects are added or taken away the total will be the same and need to re-count.
Number – order and place value MNU 0-02a	11	**On track** • The child retrieves the correct number of coins from the bag each time (five and eight). • They understand that zero means there are none and correctly give no coins. **Further learning required** • The child is unsure of the forward counting sequence and so forgets how many pieces of treasure they were asked to get. • They do not coordinate each number name with one object and instead randomly grab a handful of coins. • They do not understand that zero means there are none.

Topic – benchmarks / Es & Os	Question	Notes
Estimation and rounding MNU 0-01a	12	**On track** • The child can subitise each amount, that is, say the amount shown without counting. **Further learning required** • The child is unable to subitise up to 5 dots and gives random guesses or attempts to count the dots in ones.
Estimation and rounding MNU 0-01a	13	**On track** • The child understands the language more/less. • They subitise to make a good estimate of an amount. **Further learning required** • The child cannot answer as they do not understand the language more/less. • They make random estimates.
Estimation and rounding MNU 0-01a	14	**On track** • The child can make two collections of the same amount by matching one-to-one. **Further learning required** • The child makes two collections of different amounts. • They 'line up' the collections not paying attention to the amounts. • They do not match the objects one-to-one.
Number – addition and subtraction MNU 0-03a	15	**On track** • The child appreciates that they have to join the collections to find how many altogether. • They combine the amounts to give the correct total. **Further learning required** • The child does not understand that the two collections must be combined to find the total. • The child miscounts when combining the collections.
Number – addition and subtraction MNU 0-03a	16	**On track** • The child understands that if they 'hide' some, they will be left with fewer than they had at the start. • They can count or subitise collections of four and two coins. **Further learning required** • The child does not understand that when taking away, the amount becomes less. • They are unable to count or subitise collections of four and two coins.

Topic – benchmarks / Es & Os	Question	Notes
Fractions, decimal fractions and percentages MNU 0-07a	17	**On track** • The child can share the amount equally between two groups. • They can say how many are in each group. **Further learning required** • The child may 'share' the coins between the two pirates, giving some to each, but does not appreciate that for the shares to be fair, each pirate should be given the same number of coins.
Number and number processes MNU 0-03a	18	**On track** • The child can partition a set of six objects into two or more groups, saying how many in each group and how many altogether. **Further learning required** • The child may not trust that there are always six coins in total and need to re-count the number of coins each time.
Patterns and relationships MTH 0-13a	19	**On track** • The child understands the repetitive nature of patterns and creates a simple pattern. • The child can talk confidently about their pattern. **Further learning required** • The child does not understand the term pattern and so is unable to complete the task. • They create a design that is not a pattern.
Patterns and relationships MTH 0-13a	20	**On track** • The child can identify the missing number using the number sequence. • The child can use appropriate language when talking about the number sequence. **Further learning required** • They child has to 'drop back' and say the number sequence up to the number given to enable them to identify the number before and the number after.
Number – order and place value MNU 0-02a	21	**On track** • The child can recognise ordinal words and show this in relation to the position of the pirates; for example, first in line is number one and third is number three. **Further learning required** • The child is unable to use ordinal language correctly; for example, they may describe the third pirate as being 'three' but not 'third'.

Topic – benchmarks / Es & Os	Question	Notes
Number – order and place value MNU 0-02a	22	**On track** • The child can give one more or one less by counting on to the next number or back to the number before. **Further learning required** • The child has to count all the pirates to find the answer. • They give the wrong number.
Number – order and place value MNU 0-02a	23	**On track** • The child knows a double means twice as many/the amount is repeated. • They can show double 2, 3, 4 and 5 with concrete materials or pictures. **Further learning required** • They do not understand what double means and/or cannot work out how to solve the problem.
2D shapes and 3D objects MTH 0-16a	24	**On track** • The child can recognise *some* common 2D shapes and 3D objects. • They are beginning to use appropriate mathematical language to describe some properties of the shapes; for example, round, curved, straight, flat. **Further learning required** • The child is unable to recognise common 2D shapes and 3D objects, for example, they may call a circle a 'round'. • They do not use mathematical language such as curved, round, flat and straight.
Money MNU 0-09a	25	**On track** • The child can identify some coins. • They are beginning to show an understanding of how money is used. **Further learning required** • The child is unable to name any coins. • The child shows no awareness of how money is used.
Time MNU 0-10a	26	**On track** • The child can correctly sequence the pictures using appropriate language, for example, first and next. • They can talk about what happens at different times of the day. **Further learning required** • The child is unable to sequence the pictures. • They do not associate different events with times of the day.

Topic – benchmarks / Es & Os	Question	Notes
Measurement MNU 0-11a	27	**On track** • The child understands the language of measure and can choose the appropriate items in terms of length, mass and capacity. • The child can use appropriate measures to test their thinking. For example, the child may compare length by lining the objects up, 'weigh' the objects in their hands and use pouring to check the capacity. **Further learning required** • The child does not understand the terms longest/shortest, heaviest/lightest or holds most/holds least and so is unable to complete the task.
Angles, symmetry and transformation MTH 0-17a	28	**On track** • The child can understand and use the language of position – for example, beside, above, below, near – to locate the objects on the map. **Further learning required** • The child does not understand the language of position and so is unable to complete the task.
Angles, symmetry and transformation MTH 0-19a	29	**On track** • The child recognises symmetry as a feature and is drawn towards this. **Further learning required** • The child does not notice or refer to symmetry.
Data handling and analysis MNU 0-20a MNU 0-20b	30	**On track** • The child correctly sorts the objects by colour. • They can use the information gathered to identify which pirate has more. **Further learning required** • The child does not pay attention to the correct property to sort the coins correctly. • They cannot compare the two amounts to answer the question, 'Which pirate has more?' • They miscount the coins.
Data handling and analysis MNU 0-20a MNU 0-20b	31	**On track** • The child can identify common properties and choose one to sort the pirates. • They can explain their thinking. **Further learning required** • The child cannot find commonalities to sort the pirates. • They cannot explain how they sorted the pirates.

- Add the names of the children in the group/class to the top of columns 3-12. Use multiple sheets depending on the size of your class.
- Mark each child as they complete their Early Level Yearly Progress Check. Suggested coding is:

A. Chose an appropriate strategy/method and used it correctly

B. Chose an appropriate strategy/method but used it incorrectly or made an error in calculation

C. Chose an inappropriate strategy/method or did not attempt the question

Question	Domain											
1	MNU 0-02a											
2	MNU 0-02a											
3	MNU 0-02a											
4	MNU 0-02a											
5	MNU 0-02a											
6	MNU 0-02a											
7	MNU 0-02a											
8	MNU 0-02a											
9	MNU 0-02a											
10	MNU 0-02a											
11	MNU 0-02a											
12	MNU 0-01a											
13	MNU 0-01a											
14	MNU 0-01a											
15	MNU 0-03a											
16	MNU 0-03a											
17	MNU 0-07a											
18	MNU 0-03a											
19	MTH 0-13a											
20	MNU 0-02a											
21	MNU 0-02a											
22	MNU 0-03a											
23	MNU 0-03a											
24	MTH 0-16a											
25	MNU 0-09a											
26	MNU 0-10a											
27	MNU 0-11a											
28	MTH 0-17a											
29	MTH 0-19a											
30	MNU 0-20a MNU 0-20b											
31	MNU 0-20a MNU 0-20b											

Teacher instructions

Context

Explain to the children they are going on a space adventure with lots of maths problems to solve. The questions can be done in the suggested groupings or all together. Work with a small group of children at a time so that you can observe the strategies each child uses. Questions 7, 8, 10 and 21 need to be done one-to-one and recorded on the child's pupil sheet. Resource sheets are provided.

Suggested grouping of questions

1–7 • 8–12 • 13–16 • 17–18 • 19–20 • 21–23 • 24–26 • 27–29

Resources

Resources provided online:
Dot cards 1–10 • Alien pictures • Gem resource sheet

Questions

1. **Look at the numbers in the rockets. Write in the missing numbers.**

2. **Look at the numbers in the spaceships. Write in the missing numbers.**

3. **Write the following numbers in the moons – thirteen, zero, twenty, twelve, eight.**

4. **Look at the numbers on the aliens. Rewrite the numbers in order from the biggest to the smallest.**

5. **Look at the planets. Write in the missing numbers before, after or in-between the given numbers.**

6. **Count how many aliens there are, write the total in the box.**

7. **The astronaut can count up to thirty. Can you count up to thirty? Please start at fourteen.** Record whether the child recites all of the numbers from 14 to 30 in the correct order.

8. **a** Give the child 18 aliens (toys or photocopy and cut out the aliens from Resource 13 twice). **Count the aliens. How many aliens are there? Now write the total in the box.**

   b Now move the aliens into a circle. **How many aliens are there?**

9. **Count each alien's treasure and write the total in the box.** Discuss why the third alien might be looking glum. Can the child explain that zero means there are none?

10. Show the children a) some dot cards to 10 – regular dot patterns (five-wise ten-frames; pair-wise ten-frames; domino patterns) and b) irregular dot patterns (see Resources 2–5). Show each card for a couple of seconds and ask **How many dots are there? How do you know? How did you see the 7? etc.**

11. Ask the child **Look at the space coins here.** Point to pile A. **How many coins do you think there are (without counting)? Now look at these space coins.** Point to pile B. **Do you think there are more coins or less coins? Circle the correct answer.** Ask the child to explain their thinking. **Now count the coins in each pile to check your guess.** Discuss how accurate their estimates were.

12. **I will give you a number problem. The astronaut saw eight spaceships. Three flew away. How many spaceships are there now? Write a number sentence to show how you worked it out.** The child can choose how to solve the problem, e.g. using concrete materials, drawings, number facts.

13. **I will give you another number problem. There were four aliens on the planet Mars. Six more aliens came. How many aliens are there now? Write a number sentence to show how you worked it out.** The child can choose how to solve the problem, e.g. using concrete materials, drawings, number facts.

14. **There are nine seats on the rocket. Five astronauts are in the seats already. How many empty seats are there? Write a number sentence to show how you worked it out.** The child can choose how to solve the problem, e.g. using concrete materials, drawings, number facts.

15. a **Draw a line to cut the rocket in half.**

    b **There are four planets. Draw three aliens in each planet. How many aliens are there altogether?**

16. **How many different ways you can partition six? Write your answers in the space. You may use the alien pictures to help you.**

17. **Continue the pattern on the flag.**

18. **Look at the flags. Continue the number patterns.**

19. **The aliens are racing to the spaceship. Colour the second alien blue and the fourth alien orange.**

20. **Double the amount of plasma energy beads on each string. Write the new total in the box.**

21. **Alien Archie only wants jewels with curved edges. Alien Ava only wants jewels with straight edges. Draw a line to match the treasure to each alien. Do you know the shape names of any of Alien Archie's jewels? What about Alien Ava's? Which jewel does neither alien want? What can you tell me about this jewel?**

22. a **Alien Ava buys an apple for 8p. Draw the coins she could use in the box. Try to use as few coins as possible.** Have coins available for the children to choose and draw around if needed.

    b **Alien Archie buys a banana for 3p and an orange for 4p. How much does he spend? Write your answer in the box. Now draw his change from 10p in this box.**

23. **Look at the digital clock. What time does the rocket blast off from Earth? Write the answer on the line.**

    The rocket returns at eleven o'clock. Draw the hands on the analogue clock to show eleven o'clock.

24. a **Order the containers from 'holds the most' to 'holds the least' by numbering them 1–3.**

    b **Tick the moon dust which is heavier.**

    c **Order the flag poles from the tallest to the shortest by numbering them 1–3.**

25. a **Colour the object to the left of the crater blue.**

    b **Colour the object below the rocks green.**

    c **Colour the object on the top right of the treasure map red.**

26. **Draw the rest of the flag to make it symmetrical.**

27. **Complete the table. Which object is there most of? Write your answer on the line.**

28. Cut out the gems in Resource 14. **Sort the gems by colour (black, white, spotty). Put the gems into the pictogram.**

29. a **Look at the pictogram you have made and tell me which gem there is most of?**

    b **Which gem is there least of?**

    c **What is the difference between the most gems and the least gems?**

1.

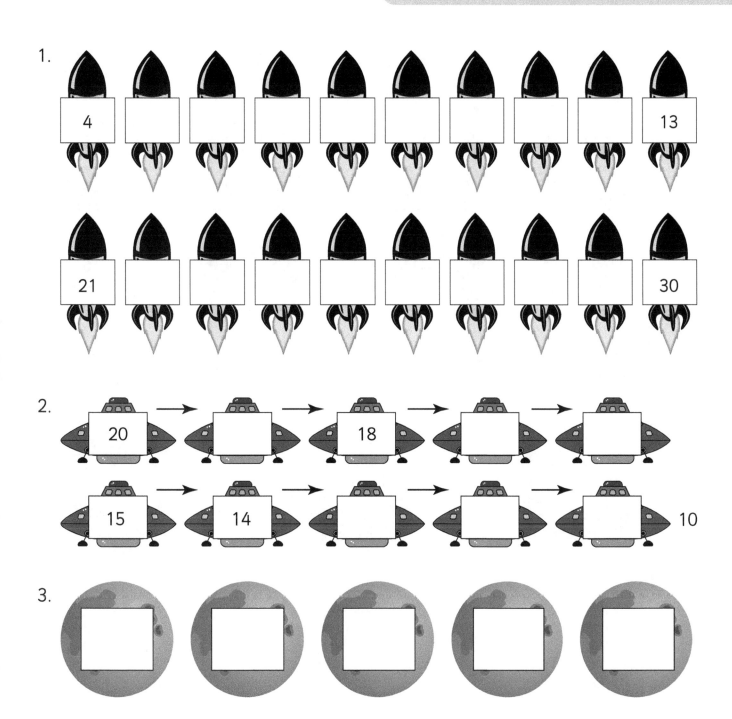

2.

3.

4.

 12

 8

 19

 5

5. **a**

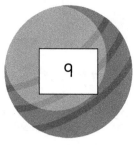 9

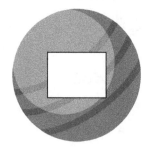 10

b

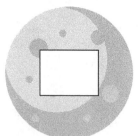

 12

 11

c

 20

 18

d

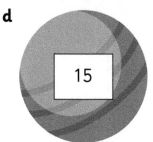 15

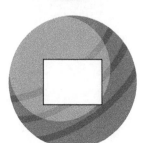

 13

6.

7.

I can count
up to 30.
Can you?

Yes / no

8. a) 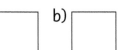 b)

Teacher comments

9.

_____

_____

_____

10. a) b)

Teacher comments

11.

Pile A

Pile B

More / Less

12.

13.

14.

15. **a**

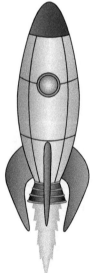

b

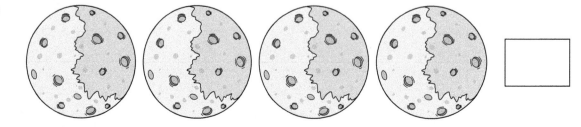

16.

17.

18.
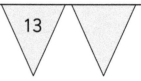

19.

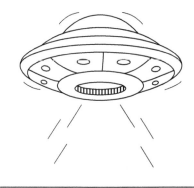

20.

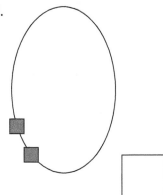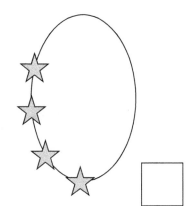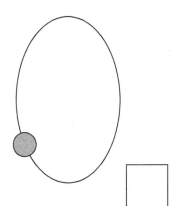

21.

Alien Archie   Alien Ava

22. **a**

b

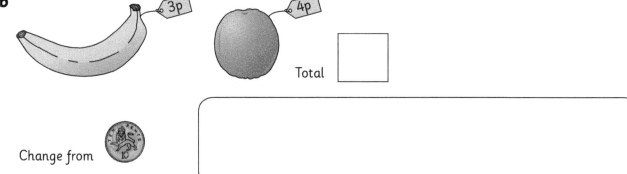

Total ☐

Change from 🪙

23.

09:00

_____

11 o'clock

24. **a**

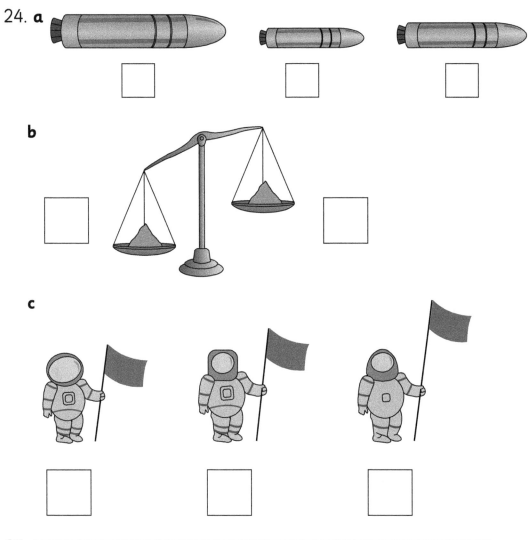

b

c

25.

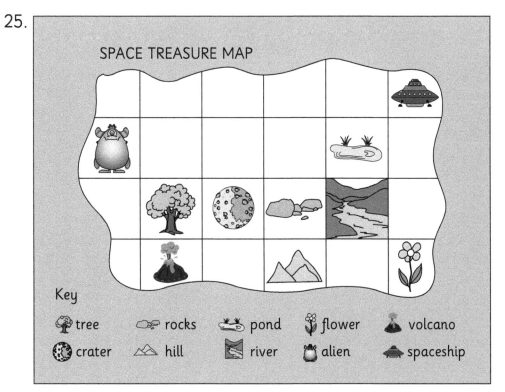

SPACE TREASURE MAP

Key

🌳 tree   🪨 rocks   🌿 pond   🌷 flower   🌋 volcano

🌑 crater   ⛰ hill   🏞 river   👽 alien   🛸 spaceship

26.

27.

	Count	Total
coins		
gems		
cups		

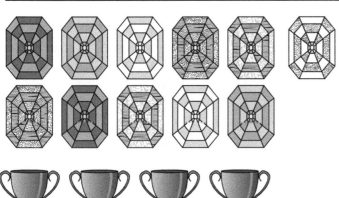

There are most ....

Name: ....

28.

Pirate Penny's Gems

6		
5		
4		
3		
2		
1		
Black	White	Spotty

29. Which has most? [ ]

Which has least? [ ]

What is the difference between the most and the least? [ ]

1. 5, 6, 7, 8, 9, 10, 11, 12 and 22, 23, 24, 25, 26, 27, 28, 29

2. 19, 18, 17, 16, 15, 14, 13, 12, 11

3. numbers correctly written in the moons – 13, 0, 20, 12, 8

4. 19, 12, 8, 5

5. **a** 11, **b** 13, **c** 19, **d** 14

6. 13

7. Delete as appropriate and note the counting sequence the pupil can correctly recite.

8. **a** 18 – note if the child can write the number 18 correctly.

   b You are looking for the child to give the answer straight away without re-counting after moving them.

9. 7, 9, 0

10. On the pupil sheet tick the box if the child can successfully subitise to 10 in each case.

11. Pile A – 14; Pile B – 19; more

12. $8 - 3 = 5$

13. $4 + 6 = 10$

14. $5 + 4 = 9$; $4 + 5 = 9$; $9 - 5 = 4$ or $9 - 4 = 5$

15. **a** The child should draw a line to cut the rocket roughly into two equal parts. Their answer should be sufficiently accurate to suggest that they understand there should be two parts of the same size.

    b 12

16. 1 and 5, 5 and 1, 2 and 4, 4 and 2, 3 and 3, 1 and 1 and 1 and 1 and 1 and 1, 2 and 2 and 2, 1 and 2 and 3, 2 and 3 and 1 or 3 and 2 and 1. Children should write a number fact to match each different way.

17. pattern continued correctly on the flag

18. 10, 11, 12, 14 and 18, 15, 14, 13

19. second alien coloured blue; fourth alien coloured orange

20. 4, 8 and 2

21. Archie – circle and oval; Ava – square, rectangle, star, pentagon, hexagon and rectangle; neither – semi-circle

22. **a** 5p, 2p, 1p – mark other combinations that make 8p as correct but note the child's ability

    b 7p and 3p (2p and 1p coins or $3 \times 1p$ coins)

23. 9 o'clock; big hand pointing straight up, little hand at eleven

24. **a** 1, 3, 2

    b tick on left

    c 3, 2, 1

25. **a** blue – tree

 b green – hill

 c red – spaceship

26. rest of the flag successfully drawn to look symmetrical

27. Table completed correctly to show 9 coins, 11 gems and 7 cups. Accept numerals or tally marks in the 'count' column.

 most – gems

28. gems sorted into black, white and spotty; correctly put into the pictogram

29. **a** most black

 b least white

 c difference 3

Topic benchmarks/ Es & Os	Question	Marking guidance
Number – order and place value MNU 0-02a	1	**On track** • The child writes each missing number correctly and completes the forward sequence accurately and confidently. **Further learning required** • The child omits a number or numbers, especially when crossing a decade or in the early teens. • They mix up the number sequence. • They reverse the digits when writing some numbers from 12 to 30; for example, the child writes 15 as 51. • They require a visual support – for example, a number track – to help them complete the task.
Number – order and place value MNU 0-02a	2	**On track** • The child writes each missing number correctly and completes the backwards sequence accurately and confidently. **Further learning required** • The child omits a number or numbers. • They mix up the number sequence. • They reverse the digits when writing some numbers from 20–10; for example, the child writes 15 as 51. • They require a visual support – for example, a number track – to complete the task.
Number – order and place value MNU 0-02a	3	**On track** • The child will immediately link each oral number word with its corresponding numeral and write it correctly. **Further learning required** • The child may be able to connect the oral number word with its symbol for one-digit numbers but find two-digit numbers more challenging, most commonly reversing digits; for example, writing 13 as 31. • They require a visual support – for example, a number track – to complete the task.
Number – order and place value MNU 0-02a	4	**On track** • The child understands that the numbers symbolise quantities and can use this knowledge to confidently order the numbers. **Further learning required** • The child regards each digit in the numbers 12 and 19 as separate and so orders the numbers as follows: 1, 1, 2, 5, 8, 9. • They disregard the 'ones' digit in 2-digit numbers and so believe that 12 and 19 have the same value and are both 'the smallest number'. • They need to use the 0–20 counting sequence to order the numbers. • They require a visual support – for example, a number track – to complete the task.

Topic benchmarks/ Es & Os	Question	Marking guidance
Number – order and place value MNU 0-02a	5	**On track** • The child can write the number before, after or in-between accurately and confidently. **Further learning required** • The child needs to say the complete forwards or backwards counting sequence to find the numbers before, after and in-between. • They require a visual support – for example, a number track – to complete the task.
Number – order and place value MNU 0-02a	6	**On track** • The child works systematically, marking each alien as it is counted. • They recognise that the number name of the last object counted also gives the total for the group. • They can write '13' correctly. **Further learning required** • The child mixes up the forward counting sequence. • They are unable to coordinate each number name with one object only, counting objects more than once or missing objects. • They write the numeral 13 incorrectly, most commonly reversing the digits.
Number – order and place value MNU 0-02a	7	**On track** • The child can start from the given number and continue the forward counting sequence up to 30, accurately and confidently. **Further learning required** • The child has to drop back to a number they feel confident with before continuing the forward counting sequence. • They omit a number or numbers, especially when crossing a decade. • They mix up the number sequence. • They require a visual support – for example, a number track – to complete the task.
Number – order and place value MNU 0-02a	8 a	**On track** • The child uses one-to-one correspondence to count the amount accurately. • They recognise that the number name of the last object counted also gives the total for the group • They write the numeral '18' correctly. **Further learning required** • The child needs to re-count to check the total amount. • They mix up the forward counting sequence. • They count an object more than once or miss an object. • They write the numeral 18 incorrectly, most commonly reversing the digits.

Topic benchmarks/ Es & Os	Question	Marking guidance
Number – order and place value MNU 0-02a	8 b	**On track** • The child understands that the amount stays the same, no matter the arrangement, and does not need to re-count the objects. **Further learning required** • The child needs to re-count to check that the total has not changed.
Number – order and place value MNU 0-02a	9	**On track** • The child accurately counts and labels the first two collections (the more able child may be able to subitise the 9 pattern). • They understand that zero means there are none and correctly label the third crater with the numeral '0'. **Further learning required** • The child mixes up the forward counting sequence. • They count an object more than once or miss an object. • They do not recognise that zero '0' represents a lack of quantity. • They are unable to use the symbol '0' for zero.
Estimation and rounding MNU 0-01a	10 a b	**On track** • The child can subitise five-wise patterns on ten-frames, using 5 and 10 as reference points. For example, instantly sees 7 as 5 and 2; 9 as 1 less than 10, etc. • They can subitise pair-wise patterns and use these to recognise odd and even numbers. For example, the child instantly sees 7 as double 3 and 1 more; 8 as double 4 or 2 less than 10, etc. **Further learning required** • The child is unable to subitise amounts to 10 and gives random guesses or attempts to count the dots in ones.
Estimation and rounding MNU 0-01a	11	**On track** • The child estimates how many coins there are in Pile A by looking and uses this information to correctly predict that Pile B has more. • They can use one-to-one correspondence to check the accuracy of their estimate/prediction, saying the number names in the correct order as they do so, and recording the amounts accurately. **Further learning required** • The child is unable to use number sense and/or visualisation to estimate and compare amounts. • They immediately attempt to count the coins in each collection or make a random, unrealistic guess.

Topic benchmarks/ Es & Os	Question	Marking guidance
Number – addition and subtraction MNU 0-03a	12	**On track** • The child recognises the problem as one which requires subtraction. • They can choose an appropriate method – for example, using concrete materials, drawing a picture, counting back on a number line or using a known fact – to find the correct answer. • They can write a calculation which mirrors their actions/ thoughts, using the appropriate symbols. **Further learning required** • The child misinterprets the problem; for example, adds rather than subtracts. • They choose the correct operation but use materials or a number line incorrectly. • They are unable to write a number sentence that mirrors how they solved the problem.
Number – addition and subtraction MNU 0-03a	13	**On track** • The child recognises the problem as one requiring addition. • They can choose an appropriate method – for example, using concrete materials or a number line, drawing a picture, or using a known fact – to find the correct answer. • They can write a calculation which mirrors their actions/ thoughts, using the appropriate symbols. **Further learning required** • The child misinterprets the problem; for example, subtracts rather than adds. • They choose the correct operation but use the number line or materials incorrectly. • They are unable to write a number sentence that mirrors how they solved the problem.
Number – addition and subtraction MNU 0-03a	14	**On track** • The child can choose how to represent and solve the problem through role-playing the situation: equalising sets of concrete materials; counting on or back on a number line; drawing a picture or part-part-whole model; using a known fact. • They can write a number sentence to match their thinking using the appropriate symbols. **Further learning required** • The child is unable to represent the problem with concrete materials or pictorially. • They can represent the problem and work out that there are four empty seats but are unable to write a number sentence to match what they did. • They write a number sentence which is incorrect; for example, they put numbers and symbols in the wrong places.

Topic benchmarks/ Es & Os	Question	Marking guidance
Fractions, decimal fractions and percentages MNU 0-07a	15 a	**On track** • The child splits the image into two roughly equal parts. **Further learning required** • The child may split the shape into two parts that are clearly unequal, referring to a 'big half' and a 'little half'. • They may split the shape into more than two parts.
Fractions, decimal fractions and percentages MNU 0-07a	15 b	**On track** • The child can form equal groups and count accurately to find the total. **Further learning required** • The child cannot form equal groups. • They may believe the total is three or four, or miscount when finding the total.
Number – addition and subtraction MNU 0-03a	16	**On track** • The child can partition a set of six objects into two or more parts, confidently recognising that the total is always the same. • They can record the appropriate number facts using numerals and symbols. **Further learning required** • The child may not trust that there are always six aliens in total and need to re-count the number of aliens each time. • They may be able to partition the set of six objects in different ways but be unable to record the corresponding number sentences.
Patterns and relationships MTH 0-13a	17	**On track** • The child can recognise and continue the pattern. **Further learning required** • The child does not understand the term 'pattern' and so is unable to complete the task. • The child draws the appropriate shapes but does not continue the pattern correctly.
Patterns and relationships MTH 0-13a	18	**On track** • The child can complete each sequence by filling in the missing numbers. **Further learning required** • The child is unable to complete the number sequence or does so incorrectly.

Topic benchmarks/ Es & Os	Question	Marking guidance
Number – order and place value MNU 0-02a	19	**On track** • The child can link the words second and fourth with the numerals 2 and 4. **Further learning required** • The child is unable to use ordinal language correctly; for example, they may describe the second alien as 'number two in the line'.
Number – addition and subtraction MNU 0-03a	20	**On track** • The child understands that 'double' means 'twice as many'. • They are able to mentally recall double facts up to double 5. • They may recognise that doubles are always even numbers. **Further learning required** • The child does not understand the request to 'double' each amount. • They may understand the meaning of the word 'double' but are unable to double each amount (that is, find $2 + 2$, $4 + 4$, $1 + 1$) mentally.
2D shapes and 3D objects MTH 0-16a	21	**On track** • The child can sort the shapes correctly using the given criteria and confidently discuss their properties using appropriate mathematical language; for example, curved, straight. • They can recognise and name common 2D shapes. **Further learning required** • The child does not understand the terms 'curved' and 'straight' and so is unable to sort the shapes correctly. • They are unable to name and/or describe common 2D shapes.
Money MNU 0-09a	22 a	**On track** • The child can choose the fewest coins possible to make the correct amount. **Further learning required** • The child makes the correct amount but uses more than three coins. • They make the amount using only pennies showing a lack of understanding of equivalent coins. • They are unable to make the correct amount.

Topic benchmarks/ Es & Os	Question	Marking guidance
Money MNU 0-09a	22 b	**On track** • The child can choose the correct calculation to find the total cost and record this using the pence symbol. • They can choose, and correctly use, a strategy for calculating change. • They can draw coins to represent the correct amount of change. **Further learning required** • The child only solves part of the problem. • They perform one or both calculations incorrectly. • They do not choose appropriate coins to show the correct answer. • They do not record the total using the pence symbol.
Time MNU 0-10a	23	**On track** • The child can confidently read and record o'clock times in analogue and digital forms. • They can talk about minutes and hours and can show which hand, or part of the digital display, is represented by each. **Further learning required** • The child writes the digital time as hour: 12, hour: 60 or hour: blank. • The child shows '11 o'clock' on the clock face but makes the hands the same length or mixes them up.
Measurement MNU 0-11a	24 a	**On track** • The child orders the containers correctly and, in discussion, shows an understanding of the terms 'most' and 'least'. **Further learning required** • The containers are ordered incorrectly. • The child is unable to use appropriate language to explain their thinking.
Measurement MNU 0-11a	24 b	**On track** • The child correctly identifies which side is heavier and can use appropriate language to describe and compare the mass of each pile of moondust. **Further learning required** • The wrong moondust is chosen as the child does not read the scales correctly. • The child is unable to use the appropriate language to explain their thinking.

Yearly progress check 2   Marking guidance

Topic benchmarks/ Es & Os	Question	Marking guidance
Measurement MNU 0-11a	24 c	**On track** • The child orders the flagpoles correctly and, in discussion, shows an understanding of the terms 'tallest' and 'shortest'. **Further learning required** • The flagpoles are ordered incorrectly. • The child is unable to use the appropriate language to explain their thinking.
Angles, symmetry and transformation MTH 0-17a	25 a b c	**On track** • The child can understand and use the language of position to locate the objects on the map. **Further learning required** • The child does not understand the language of position – for example, left, right, top, bottom, above, below – and so is unable to locate the objects correctly.
Angles, symmetry and transformation MTH 0-19a	26	**On track** • The child understands that each side is a mirror image/ reflection of the other and can complete the missing half of the circle with a reasonable degree of accuracy. **Further learning required** • The child draws the other 'half' of the flag but it is not symmetrical, i.e. it is copied but not reflected.
Data handling and analysis MNU 0-20a MNU 0-20b	27	**On track** • The child can collect information by making a mark for each object counted and complete the table correctly. • They can count the marks to find each total and use the results to answer questions, confidently. **Further learning required** • The child does not count the objects correctly, missing some or counting an object more than once, therefore not making one mark for each object. • They record results in the wrong place in the table or miscount the marks. • They are unable to answer questions about the table.
Data handling and analysis MNU 0-20a MNU 0-20b	28	**On track** • The child can correctly sort the objects by colour/pattern and record their results in the pictogram. **Further learning required** • The child does not pay attention to the correct property to sort the objects correctly. • They do not display the objects correctly in the pictogram.

Topic benchmarks/ Es & Os	Question	Marking guidance
Data handling and analysis MNU 0-20a MNU 0-20b	29 a b c	**On track** • The child understands the vocabulary 'most', 'least' and 'difference' and can accurately count and compare amounts to answer questions about the pictogram. **Further learning required** • The child does not understand the vocabulary, or requirement of the task, and so is unable to answer correctly. • They miscount some objects and so give an incorrect answer.

Early level yearly progress check 2   Record sheet

- Add the names of the children in the group/class to the top of columns 3-12. Use multiple sheets depending on the size of your class.
- Mark each child as they complete their Early Level Yearly Progress Check. Suggested coding is:

   A. Chose an appropriate strategy/method and used it correctly

   B. Chose an appropriate strategy/method but used it incorrectly or made an error in calculation

   C. Chose an inappropriate strategy/method or did not attempt the question

Question	Domain										
1	MNU 0-02a										
2	MNU 0-02a										
3	MNU 0-02a										
4	MNU 0-02a										
5	MNU 0-02a										
6	MNU 0-02a										
7	MNU 0-02a										
8a	MNU 0-02a										
8b	MNU 0-02a										
9	MNU 0-02a										
10a	MNU 0-01a										
10b	MNU 0-01a										
11	MNU 0-01a										
12	MNU 0-03a										
13	MNU 0-03a										
14	MNU 0-03a										
15a	MNU 0-07a										
15b	MNU 0-03a										
16	MNU 0-03a										
17	MTH 0-13a										
18	MTH 0-13a										
19	MNU 0-02a										
20	MNU 0-03a										
21	MTH 0-16a										
22a	MNU 0-09a										
22b	MNU 0-09a										
23	MNU 0-10a										
24a	MNU 0-11a										
24b	MNU 0-11a										
24c	MNU 0-11a										
25a, b, c	MTH 0-17a										
26	MTH 0-19a										
27	MNU 0-20a MNU 0-20b										
28	MNU 0-20a MNU 0-20b										
29a, b, c	MNU 0-20a MNU 0-20b										

End of early level assessment 1

Dinosaurs

Context

Investigating Dinosaurs is a very popular topic in the Early Years. This assessment is suitable for use with children who have carried out class-based topic work relating to dinosaurs. Resources and materials previously used within the topic (concrete, pictorial and in story form) will provide familiar contexts and enable children to demonstrate their understanding and apply their knowledge. However, some of the tasks could be used without prior knowledge of dinosaurs.

Resources
Resources provided online: A range of 2D shapes · Dinosaur length tables · Dinosaur wrapping paper · Dinosaur images · Coins · Dinosaurs with price tags **Resources needed and available in the setting:** 3D toy dinosaurs · Selection of fiction and non-fiction books pertaining to dinosaurs · Word bank of dinosaur names

Tasks	Es & Os	Benchmarks
Task 1 Dinosaur sort	MNU 0-20b I can match objects, and sort using my own criteria, sharing my ideas with others.	• Uses knowledge of colour, shape, size and other properties to match and sort items in a variety of different ways.
Task 2 Describe and design a dinosaur	MNU 0-02a I have explored numbers, understanding that they repesent quantities, and I can use them to count, create sequences and describe order. MTH 0-16a I enjoy investigating objects and shapes and can sort, describe and be creative with them.	• Uses one-to-one correspondence to count a given number of objects to 20. • Recognises, describes and sorts common 2D shapes and 3D objects according to various criteria; for example, straight, round, flat and curved.
Task 3 Dinosaur comparison	MNU 0-02a I have explored numbers, understanding that they represent quantities, and I can use them to count, create sequences and describe order. MNU 0-11a I have experimented with everyday items as units of measure to investigate and compare sizes and amounts in my environment, sharing my findings with others. MNU 0-20a I can collect objects and ask questions to gather information, organising and displaying my findings in different ways.	• Orders all numbers forwards and backwards within the range 0–20. • Compares and describes lengths, heights, mass and capacities using everyday language including longer, shorter, taller, heavier, more and less. • Estimates, then measures, the length, height, mass and capacity of familiar objects using a range of appropriate non-standard units. • Interprets simple graphs, charts and signs and demonstrates how they support planning, choices and decision-making.

End of early level assessment 1

Tasks	Es & Os	Benchmarks
Task 4a, 4b and 4c Estimation	MNU 0-01a I am developing a sense of size and amount by observing, exploring, using and communicating with others.	• Recognises the number of objects in a group, without counting (subitising). • Checks estimates by counting. • Demonstrates skills of estimation in the contexts of number and measure using relevant vocabulary, including less than, longer than, more than and the same.
Task 5a and 5b Pattern	MTH 0-13a I have spotted and explored patterns in my own and the wider environment and can copy and continue these and create my own patterns.	• Copies, continues and creates simple patterns involving objects, shapes and numbers. • Explores, recognises and continues simple number patterns.
Task 6 Sharing	MNU 0-07a I can share out a group of items by making smaller groups and can split a whole object into smaller parts.	• Shares out a group of items equally into smaller groups.
Task 7 Measurement Angle, symmetry and transformation	MNU 0-11a I have experimented with everyday items of measure to investigate and compare sizes and amounts in my environment, sharing my findings with others. MTH 0-17a MTH 0-19a I can use simple directions and describe positions. I have fun creating a range of symmetrical pictures and patterns using a range of media.	• Compares and describes lengths, heights, mass and capacities using everyday language, including longer, shorter, taller, heavier, lighter, more and less. • Estimates, then measures, the length, height, mass and capacity of familiar objects using a range of appropriate non-standard units. • Understands and correctly uses the language of position and direction, including in front, behind, above, below, left, right, forwards and backwards. • Identifies, describes and creates symmetrical pictures with one line of symmetry.
Task 8 Money	MNU 0-09a I am developing my awareness of how money is used and can recognise and use a range of coins.	• Identifies all coins to £2. • Applies addition and subtraction skills and uses 1p, 2p, 5p and 10p coins to pay the exact value for items to 10p.

End of early level assessment 1

TASK 1

Dinosaur sort

- Harry got his bucket of dinosaurs out of his toy box. To his dismay they were all mixed up! **Can you sort them for Harry? How did you sort them? Can you sort them another way?**

- Provide the children with a box or bucket containing 3D toy dinosaurs. The children should be familiar with the different types of dinosaurs and be able to use appropriate vocabulary to describe them.

- Now, after reviewing the children's suggestions, sort the dinosaurs using a different criterion and ask the children to tell you what you did to sort the dinosaurs. For example, the dinosaurs could be sorted by those standing on two legs and those standing on four legs. **How did I sort the dinosaurs?** Observe whether the children can apply their knowledge of sorting to describe the criterion used.

TASK 2

Describe and design a dinosaur

- Provide the children with a range of 2D shapes (or see Resource 15) and ask them to make their own dinosaur using the shapes provided.

- **What can you tell me about the shapes you have used? What do they represent?** Check what properties and facts the pupils can tell you about each shape.

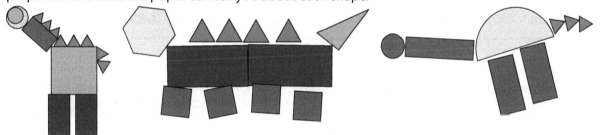

Design a dinosaur

- Provide the children with scissors, gummed paper and a plain A4 or A3 sheet of paper. Ask the children to create their own dinosaur. Tell them that their dinosaur can *only* be made using *triangles*, *squares* and *rectangles* and that they have to cut their own shapes before they can compose their picture. For children unable to cut, pre-cut shapes can be provided.

- **What shapes did you use to create your dinosaur? What can you tell me about the shapes you have used? How many triangles, squares and/or rectangles did you use?** For example, small triangles for teeth or big triangles on the dinosaur's back.

TASK 3

Dinosaur comparison

- **Harry has the following dinosaurs. Their lengths are shown in the table** (Resource 16a or 16b). Provide the children with a copy of the table: this may also be written on a whiteboard or produced on a smart board. Choose a non-standard unit that 'Length' in the tables refers to; for example, cubes or counters. Children can be shown either Table a or b depending on their readiness to handle different ranges of numbers.

- **Put the dinosaurs in order by length.** Have prepared a number line with 0 at one end and 30 at the other. Give each child a copy of the number line to assist putting the dinosaurs in order. See the example below.

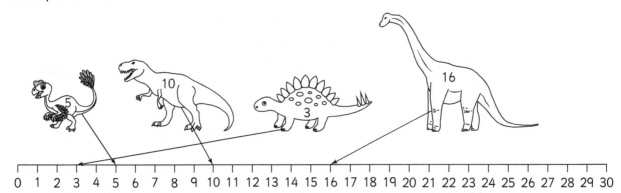

- Question the children about the position of the dinosaurs. **How did you order the dinosaurs? Which one is the shortest/longest? Are any dinosaurs the same length? Which dinosaur is longer than the Stegosaurus? Which dinosaurs are shorter than the Iguanodon?** etc.

- Use non-fiction books to further explore size. Ask the children to estimate the length/height of the animals in the pictures then find the actual lengths/heights using a chosen non-standard unit. **Can you put the dinosaurs in order from shortest to longest/tallest? Where will these numbers go on the number line?** If doing the activity practically, ask the children to place the dinosaur picture or toy dinosaur in the correct position on the number line.

- Note that although the benchmarks suggest 0–20 for identifying and recognising numbers, if the child successfully completes the task, it is good to provide a challenge by including numbers over 20.

TASK 4

Estimation

a) Dinosaur flash

- Set out some toy dinosaurs (up to 6 initially) and cover them (with a cloth). 'Flash' this collection for 2–3 seconds by removing and replacing the cloth. The children should be able to subitise these amounts. Ask **How many dinosaurs did you see? How did you see them?** Encourage use of language such as 'I saw three in a group, then one and another one'. Next, 'flash' a larger collection which is challenging to subitise. Encourage the children to use what they know about the first collection to estimate the number of dinosaurs in the new collection. **Did you see more/less dinosaurs this time? How many dinosaurs do your think there are?** Ask the children to check their estimates by counting.

End of early level assessment 1

b) Dinosaur wrapping paper

- Hold up a piece of dinosaur wrapping paper (see Resource 17). Ask the children to look at the paper carefully and explain that you are going to hide it and then ask them questions about what they saw. For example, **Can you estimate how many Tyrannosaurus rex dinosaurs you saw? How many big dinosaurs were there?** etc.

- Progress to 'flashing' the wrapping paper (holding the paper up for approximately 5 seconds) so that the children have to make a quick estimate. **How many dinosaurs are walking on two legs? How many dinosaurs have Christmas hats on? Are there more small dinosaurs with hats on than large dinosaurs? How many parcels are there?**

- The same wrapping paper can be used several times, the focus being altered by varying the questions asked.

c) Dinosaur pictures

- Show the children a dinosaur picture (as here and in Resource 18). Ask them to estimate how many dinosaurs they can see. **How many are flying? Are there more in the air or on the ground? How many have long necks? How many are in the water? Are there less in the water than flying in the air?** Progress to flashing the picture as in part b. Initially use the same picture, just alter the questions, before progressing to a different picture.

TASK 5

Pattern

a) Shape patterns

- Harry has been listening carefully and knows that not all dinosaurs are the same. Some of their differences relate to what they eat and where they live. Some dinosaurs have sharp teeth. Look at illustrations in books, both fiction and non-fiction, and online to provide children with ideas and visual experiences.

- Ask the children to **Make an AB repeating pattern**. They may choose to use triangles or other 2D shapes (see Resource 15) to depict sharp dinosaur teeth. Offering a free selection of shapes is part of the assessment. Look for the child being able to justify their choices.

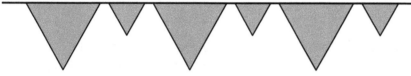

- Children can be asked to make different patterns as appropriate and to describe them. For example, big tooth, small tooth, big tooth, etc.

- Similarly, the children could use pentagons/triangles to represent bony plates and spikes on the dinosaur's back. If the child is able to, ask them to **Draw the pattern on to this picture** (see Resource 19).

- Ask the child to use shapes to represent the bony plates which give the dinosaur protection. **Can you make an AAB pattern?** Ask the child to explain/describe their representations.

b) Number patterns

- This dinosaur has three claws on each foot. It has one tusk and two horns (see Resource 20). **How many claws would there be if we had two dinosaurs? How many tusks would there be? How many horns would there be for three dinosaurs?** etc. Roll out some dough and press a toy dinosaur's footprints into it. Ask the children to decide **Which of the dinosaurs made the footprints? Now, can you repeat the pattern? Tell me about your pattern. How did you count the claws?**

TASK 6

Sharing

- **Harry knows that herbivores like to eat plants, ferns, mosses, leaves, twigs and seeds. Stegosauruses particularly like to eat ferns.**

Five Stegosaurus dinosaurs were at the swamp looking for some juicy ferns. There were 11 fern plants growing there. **How many ferns would they get each? Would they all get the same number of fern plants?** Have to hand 11 leaf-shaped pieces of paper and, if possible, five Stegosaurus dinosaur toys or pictures (if toys/pictures are unavailable, use other objects, for example counters, to represent the dinosaurs) so that the child can physically carry out the sharing exercise.

TASK 7

Measure

- Ask each child to **Choose a dinosaur.** Ask **Can you think of a way to compare the height of your dinosaur with someone else's?** If required, suggest that they could use interlocking cubes to build a tower the same height as their dinosaur and then compare the 'towers' of cubes with the other children to ascertain which is tallest/the same height/shortest, etc.
- Ask the children to **Use wooden blocks to make a cave door for your dinosaur. It has to be tall enough for the dinosaur to fit in.** Do the children use the tower of cubes to help with the height of the cave entrance? Do they build symmetrical sides to their archways? How do they measure the width of an archway? Do they estimate or use the dinosaur to check?

TASK 8

Money

- Harry wants some more dinosaurs for his bucket. He visits the local charity shop and finds a selection of dinosaurs individually priced (see Resource 22). Harry has the following coins in his wallet (see Resource 21).

- Ask the children to identify the coins. **How much money does Harry have to spend? Which dinosaurs could Harry buy? How many could he buy? Which coins will he use? Will he get any change?**

End of early level assessment 2

The party

Context

This assessment will allow information linked to days/weeks/months to be contextualised and children will have opportunities to apply their knowledge. Some preparatory work will be required if children are not familiar with a traditional calendar and its layout. Children should know how many days of the week/months of the year there are and be able to recite them. They should know that there are four designated seasons linked to our calendar.

Resources

Resources provided online:

Cupcake • Clock templates • Coins • Toy price table

Resources needed and available in the setting:

Range of calendars, diaries • Soft toy bear or equivalent • Birthday chart to be completed showing each child's birth month • Examples of clocks and other ways of measuring time

Other resources:

Refer to Chapter 6 (Time) in Early Level Maths Teacher Guide where appropriate vocabulary, games, rhymes, songs and ideas can be found.

The Party

Tasks	Es and Os	Benchmarks
Task 1 Happy Birthday, Teddy	MNU 0-10a I am aware of how routines and events in my world link with times and seasons, and have explored ways to record and display these using clocks, calendars and other methods. MNU 0-20a I can collect objects, and ask questions to gather information, organising and displaying my findings in different ways.	• Links daily routines and personal events to time sequences. • Names the days of the week in sequence, knows the months of the year and talks about features of the four seasons in relevant contexts. • Asks simple questions to collect data for a specific purpose. • Applies counting skills to ask and answer questions and makes relevant choices and decisions based on the data. • Contributes to concrete or pictorial displays where one object or drawing represents one data value, using digital technologies as appropriate. • Interprets simple graphs, charts and signs and demonstrates how they support planning, choices and decision-making.

End of early level assessment 2

Tasks	Es and Os	Benchmarks
Task 2 a) Invitations b) Making a clock	MNU 0-02a I have explored numbers, understanding that they represent quantities, and I can use them to count, create sequences and describe order. MNU 0-10a I am aware of how routines and events in my world link with times and seasons, and have explored ways to record and display these using clocks, calendars and other methods.	• Recalls the number sequence forwards within the range 0-30, from any given number. • Identifies and recognises numbers from 0-20 • Recognises, talks about and where appropriate, engages with everyday devices used to measure or display time, including clocks, calendars, sand timers and visual timetables. • Reads analogue and digital o'clock times (12 hours only) and represents this on a digital display or clock face. • Uses appropriate language when discussing time, including before, after, o'clock, hour hand and minute hand.
Task 3 Planning a party	MNU 0-02a I have explored numbers, understanding that they represent quantities, and I can use them to count, create sequences and describe order. MNU 0-07a I can share out a group of items by making smaller groups and can split a whole object into smaller parts.	• Uses one-to-one correspondence to count a given number of objects to 20. • Shares out a group of items equally into smaller parts.
Task 4 Birthday money	MNU 0-03a I can use practical materials and can 'count on and back' to help me understand addition and subtraction, recording my ideas and solutions in different ways. MNU 0-09a I am developing my awareness of how money is used and can recognise and use a range of coins.	• Adds and subtracts mentally to 10 • Uses appropriately the mathematical symbols +, - and = • Identifies all coins to £2 • Applies addition and subtraction skills and uses 1p, 2p, 5p and 10p coins to pay the exact value for items to 10p.

In preparation for this assessment look at a range of simple calendars to get a flavour of the information contained within them. There are lots of lovely calendars utilising cartoon characters, with an emphasis on recording family events and appointments. Do the children know how to use a calendar? Many children may use information from an iPad or a parent's phone to get the date. Look at the layout for each month. Diaries can also give a different view of weeks and months and are worth exploring. Old diaries may be present on a writing table where the children can use them informally. Children often do a daily day of the week/ weather task. Ideally, this should be seen in conjunction with the rest of the week/month rather than removing or replacing information at the beginning of the next day.

Set the scene – Bring in a soft toy bear. Tell the children it is going to be Teddy's birthday soon. He is going to be 6. He is having a party and we need to help him plan for it. His party is on the 5th May. This year 5th

End of early level assessment 2

May is a Tuesday. Teddy was born in the month of the year which is in the season of Spring. You should use your discretion to identify the most appropriate time to introduce this information. Some children will automatically make connections between this information and, for example, the singing of songs related to the days of the week/months of the year/ seasons and the reading of appropriate rhymes and stories. Others may need the connections and links made more visual. Revisit this when carrying out the tasks to help the children contextualise the information.

TASK 1

Happy birthday, Teddy

- Tell the children we now know when Teddy is going to be 6. It would be nice for Teddy to know when the boys and girls in the class had birthdays. Ask **What date were you born on?**

- Collect whole-class information on the children's birthdays. Provide each child with a symbol – for example, a balloon. Ask them to write their own name on the symbol, explaining that we are going to make a birthday/birth month display. Tell the children that together they are going to use the data gathered to make a pictogram/bar chart of their birthday months/months in which they were born. **Say and order the months of the year together.** Ask **How many months are there? Let's say them together. Which month comes first? Last? Which one is second?** Provide cards with the months of the year on them and you and the children will together order them to form the horizontal axis of the pictogram. Encourage children to put their birthday symbol above the correct month to form the columns of the pictogram.

- Can they use this information to answer some questions?

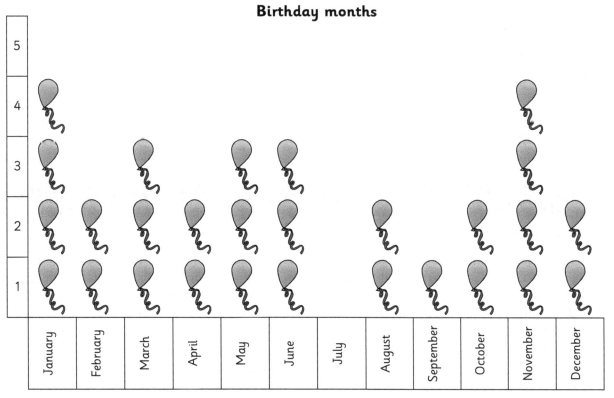

Birthday months

Ask **Who was born in the same month as Teddy? Is there a month with lots of birthdays? Which month has the most number of birthdays? How many are there in that month? Are there any months which have no birthday symbols?**

End of early level assessment 2

- Now that each child's birthday is known, use a calendar to see what day of the week the children will celebrate/celebrated their birthday on this year. Ask **How many days of the week are there? Which day comes before Wednesday? After Friday? Who has a birthday on Monday this year?** etc. The data gathered can be used to look for the most common day of the week to be born on.

- Provide the children with a picture of a cake to write their name on (see Resource 23).

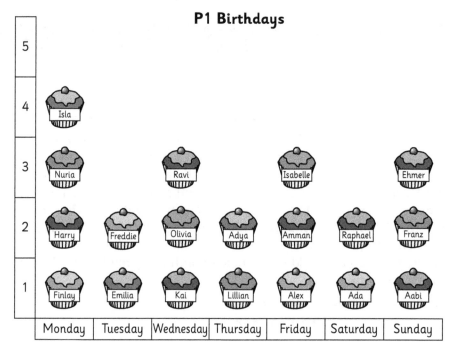

P1 Birthdays

Ask **Can you make a pictogram like Finlay's one using our group/class information?** Can the children now apply their knowledge and construct a bar chart/pictogram, similar to the whole-class, months-of-the-year task and interrogate the information visible to them? Ask appropriate questions to guide them. For example, **Who was born on a Wednesday? How many children were born on a Friday? Which was the most popular day to be born on?** Your questioning will ultimately depend on the data display.

TASK 2

a) Invitations

- Teddy wants to invite his friends to his party. He is planning to send out invitations. He wants his party to be on a Saturday. The party will start at two o'clock and finish at four o'clock. (Writing the invitations creates links with Literacy.) Ask the children **What day comes before Saturday? What day comes after Saturday?** Do the children know that Saturday and Sunday together are known as the weekend?

  Using individual teaching clocks (see Resource 24), ask the children to show two o'clock and then four o'clock. Ask the children **How long will the party last for?** Provide the children with a laminated template for both digital and analogue time (see below). Ask them to write two o'clock as a digital time and to draw the hands on a clock to show two o'clock. Ask **Can you tell me about the information you have marked on your wipe clean board?** If needed, prompt the children to describe the clock hands to you. Have they managed to draw one long hand and one short hand? **What number is the long hand pointing to? What does this mean? What can you tell me about the short hand?**

- Ask the children to wipe clean their clock template and depict four o'clock both as analogue and digital time. Again, can they describe the hands to you? Have they changed the 2 to a 4 for digital time?

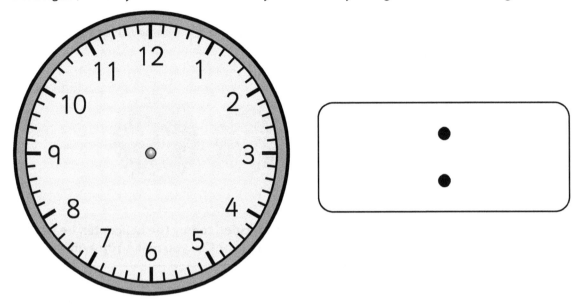

b) Making an analogue clock

- Challenge the children to make an analogue clock of their own. This assessment should not be rushed so consider setting aside an afternoon to complete it.

- The children have to design their clock. Provide scissors, a selection of paper, pens, glue sticks and some fiction and non-fiction stories related to time to inspire them. Your school library should have an appropriate selection to use for reference. The children should be able to use scissors appropriately to cut the necessary clock face and make hands. For children who find this challenging, it may be necessary to provide a range of pre-cut paper shapes that they can choose from. Ask **What shape do you think the clock face could be? How big would it need to be? What could you use to make clock hands? Can you write the numbers in order round the clock face? What number is at the top of the clock face? Can you set the hands on the clock to show either two o'clock or four o'clock?**

TASK 3

Planning the party

- Teddy has five friends he would like to invite to his party. Ask **How many chairs/plates/ cups would be needed for the party table?** (Will they remember to add Teddy himself to the count?) Ask the children to draw a sketch or make a table plan. Make concrete materials available for children who lack confidence with drawing. A range of loose parts materials would be ideal. This is a useful way of highlighting whether the children can visualise the situation and is an indicator of their understanding of one-to-one correspondence. Ask **If everyone is to get two sandwiches each, how many sandwiches do we need?** The children can then use their plan to help them decide on the sandwich numbers. Mummy Bear has baked cakes. She made 14 cakes. Mummy Bear thinks that everyone should get two cakes each. Ask **Are there enough cakes?** If required, provide concrete materials so that the children can physically share out the objects to check their answer. Alternatively, the children could further develop the use of their sketch/plan.

End of early level assessment 2

Shopping

- Teddy has some birthday money. He visits his local toy shop and spots some things that he would like to buy.
- He has the following coins (see also Resource 21).

Ask **Can you tell me what value these coins have?**

- Look at Teddy's choices (see Resource 25). Ask **If Teddy decided to buy the helicopter, he would need 10p. Can you make 10p using the coins that you have? Can you make 10p a different way?**

Toy	Cost	
	10p	
	7p	
	9p	
	3p	

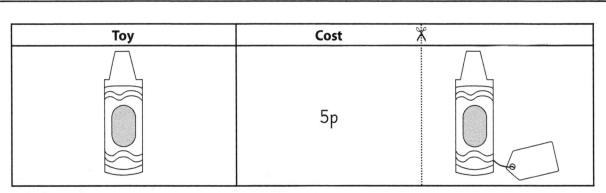

Toy	Cost	✂
	5p	

- Ask **Which three toys do you think Teddy could buy? How much would they cost altogether?** Depending on the child's choice **Would Teddy get change?**

- **If Teddy bought the bricks and the ball, how much would that cost? What coin/s would he use to pay with?**

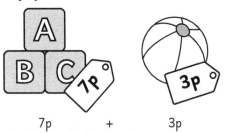

7p   +   3p   =

If Teddy bought the jeep and the crayon what coin or coins would he use to pay for them? Would he get change? Allow the children to explore their thinking using real coins if possible (or refer to Resource 21).

9p   +   5p   =

- Ask **Can you identify these coins?** (See also Resource 21.)

End of early level assessment 3

Playing games

Context

Board games play an important part in encouraging and developing children's early mathematical and numerical skills. It is vital that all children experience playing board games in a physical sense, not just a virtual sense.

Games are a great way to assess children's knowledge of numeracy/mathematics and whether they can apply their knowledge in different contexts.

Resources

Resources provided online:

Board game templates • Sudoku objects • Sudoku grid • Wet Sponge Splosh target and graph • Hungry mouse grid • Coins

Resources needed and available in the setting:

Magnetic fishing game • Bean bags or sponges • Chalk • paper/plastic plates • dice • 3D blocks • sand timer

---

Tasks	Es&Os	Benchmarks
Task 1 Board games	MNU 0-01a I am developing a sense of size and amount by observing, exploring, using and communicating with others about things in the world around me.	• Recognises the number of objects in a group, without counting (subitising).
	MNU 0-02a I have explored numbers, understanding that they represent quantities, and I can use them to count, create sequences and describe order.	• Recalls the number sequence forwards within the range 0-30, from any given number.
	MNU 0-03a I can use practical materials and can 'count on and back' to help me understand addition and subtraction, recording my ideas and solutions in different ways.	• Uses the language of before, after and in-between. • Counts on and back in ones to add and subtract.
Task 2 Sudoku	MTH 0-13a I have spotted and explored patterns in my own and the wider environment and can copy and continue these and create my own patterns.	• Copies, continues and creates simple patterns involving objects, shapes and numbers. • Explores, recognises and continues simple number patterns.
Task 3 Magnetic fishing	MNU 0-03a I use practical materials and can count on and back to help me understand addition and subtraction, recording my ideas and solutions in different ways.	• Counts on and back in ones to add and subtract. • Adds and subtracts mentally to 10.

Tasks	Es&Os	Benchmarks
	MNU 0-20a I can collect objects and ask questions to gather information, organising and displaying my findings in different ways. MNU 0-20c I can use the signs and charts around me for information, helping me plan and make choices and decisions in my daily life.	• Applies counting skills to ask and answer questions and makes relevant choices and decisions based on the data. • Contributes to concrete or pictorial displays where one object or drawing represents one data value, using digital technologies as appropriate. • Interprets simple graphs, charts and signs and demonstrates how they support planning, choices and decision-making.
Task 4 Wet sponge splosh	MNU 0-20a I can collect objects and ask questions to gather information, organising and displaying my findings in different ways.	• Interprets simple graphs, charts and signs and demonstrates how they support planning, choices and decision making. • Applies counting skills to ask and answer questions and makes relevant choices and decisions based on data.
Task 5 Hungry mouse	MTH 0-17a In movement, games and technology I can use simple directions and describe positions.	• Understands and correctly uses the language of position and direction, including front, behind, above, below, left, right, forwards and backwards to solve simple problems in movement games.
Task 6 Taking coins	MNU 0-01a I am developing a sense of size and amount by observing, exploring, using and communicating with others about things in the world around me. MNU 0-02a I have explored numbers, understanding that they represent quantities, and I can use them to count, create sequences and describe order. MNU 0-09a I am developing my awareness of how money is used and can recognise and use a range of coins.	• Checks estimates by counting. • Demonstrates skills of estimation in the contexts of number and measure using relevant vocabulary, including less than, longer than, more than and the same. • Identifies how many regular dot patterns, for example, arrays, five frames, ten frames, dice and irregular dot patterns, without having to count (subitising). • Applies addition and subtraction skills and uses 1p, 2p, 5p and 10p coins.
Task 7 Speed building	MNU 0-01a I am developing a sense of size and amount by observing, exploring, using and communicating with others about things in the world around me. MTH 0-16a I enjoy investigating objects and shapes and can sort, describe and be creative with them.	• Demonstrates skills of estimation in the contexts of number and measure using relevant vocabulary, including less than, longer than, more than and the same. • Recognises, describes and sorts common 2D shapes and 3D objects according to various criteria, for example, straight, round, flat and curved.

TASK I

Game I – Board games

a)

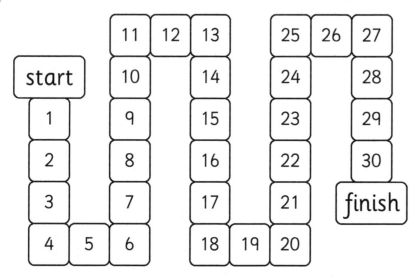

Amman and Nuria were enjoying playing a board game on the carpet (see Resource 26). The doorbell rang and Nuria's pet dog ran across the board, scattering the counters. Can you use the information here (point to the table) to find out who was in the lead and what number their counter was on?

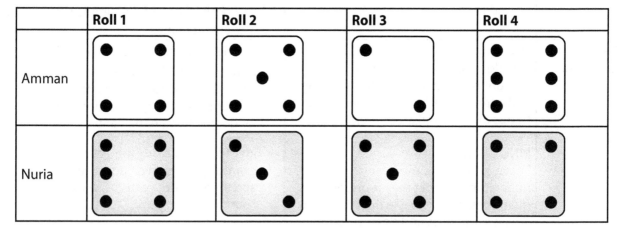

	Roll 1	Roll 2	Roll 3	Roll 4
Amman				
Nuria				

b)

10	11	12	13	14 Jump back 2	15	16 Move on 1	17	18 Jump back 3	19	20 Miss a turn
	(F)			(I)			(N)			

Isla, Finlay and Nuria were playing a different game (see Resource 26). Isla landed on 14. **What number did she have to move her counter to?** Note that you may need to read the instruction 'Jump back 2' out loud.

- Finlay's counter was on 11. He rolled a 5. **What number did he land on?**
- Nuria had her counter on 17. She rolled a one. **Which number did she finish on?**
- The board game instructions may be read aloud to the children if required.

TASK 2

Sudoku (Early)

- This is a basic example of an early Sudoku game (see Resources 27 and 28).
- Rules of Sudoku: The children should have access to three lots of three objects. They can be concrete or pictorial. They must try to ensure that each row/column in the grid only has one of each object in it.

- The objects can then be replaced by numbers, for example, 1, 2, 3. The same rules apply.

1	2	3
2	3	1
3	1	2

- Encourage the children to solve the Sudoku problem.
- This idea can then be expanded to a 4 × 4 grid, a 6 × 6 grid and progress to a 9 × 9 grid to challenge the most able.

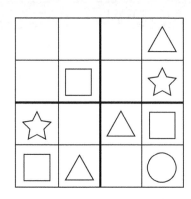

		△
	□	☆
☆	△	□
□	△	○

4	3		
1	2	3	
		2	
2	1		

TASK 3

Magnetic fishing

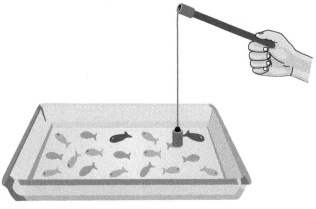

- If possible, play a magnetic fishing game with the children in advance of this assessment. Provide the children with access to a range of manipulatives, that they are used to handling, to support their investigation.
- Finlay, Nuria, Isla and Amman have been playing with their magnetic fishing game. They each had three turns. The table below (also see Resource 29) shows the children's scores for each turn.

Player	Turn one	Turn two	Turn three
Finlay	3	4	2
Nuria	5	1	5
Isla	2	2	4
Amman	0	6	2

- Say **Add up the scores for each child to find out who won the game.** Ask **What was Finlay's/ Nuria's/Isla's/Amman's total score? Who had the highest score for turn three? Who had the lowest score for turn two? Who had the same scores for turn three?**
- Encourage the children to play a magnetic fishing game of their own. Can they record their own data and decide who won their game?

End of early level assessment 3

Wet sponge splosh

- Children may enjoy playing Wet Sponge Splosh outdoors, with wet sponges and a chalked target on the playground. In this task, however, they are looking at the scores achieved by Finlay (F), Isla (I), Nuria (N) and Amman (A).

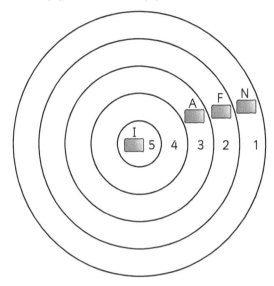

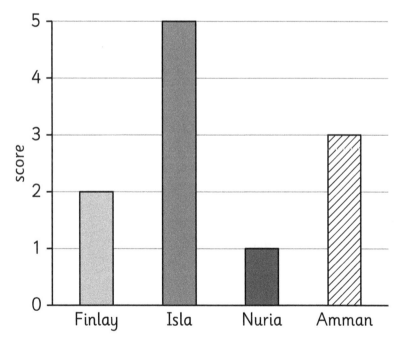

- Look at the game picture and the graph of the results with the children (also see Resource 30). Can they answer the following questions?

  1. **Who had the highest score?**

  2. **Who had the lowest score?**

  3. **Did anyone get the top score?**

  4. **What was the difference between Finlay's score and Isla's score?**

  5. **What was the difference between Nuria's score and Amman's score?**

End of early level assessment 3

TASK 5

Hungry mouse

• Mouse is trying to find his way through the maze (see Resource 31). He needs the children's help to guide him to his mouse hole and his bit of cheese. Ask **Can you help Mouse by telling him how many steps he needs to take and which direction he needs to turn – left or right?**

• Masking tape indoors or chalked grids outdoors can provide useful experience in physically following directions prior to carrying out the assessment.

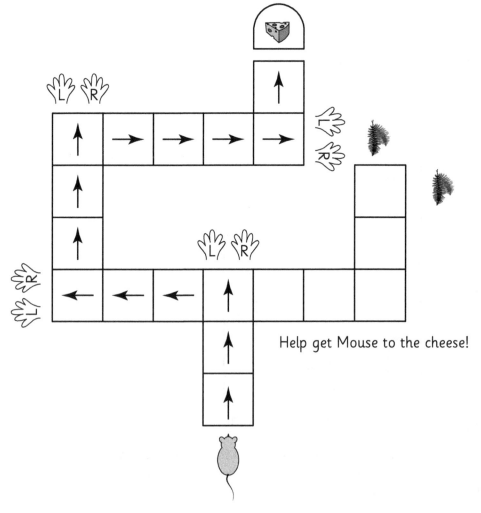

Help get Mouse to the cheese!

• **How many steps forward did Mouse take altogether? How many times did Mouse turn left? How many times did Mouse turn right?**

End of early level assessment 3

TASK 6

Taking coins

- For this game you will need one paper/plastic plate per child, 10 × 1p coins per child, a dice. The game is best played with five or six children. The children should be seated on the floor in a circle or at a table to play this game.

- Each child is given ten 1p coins to place on their plate. First, Child A rolls the dice. The number they roll on their dice decides how many 1p coins they can take from the person on their right to place on their own plate. For example, Child A rolls a 3 and takes three 1p coins from the person seated to their right. Child A now has 13p in total.

- Play continues with Child A passing the dice to Child B. Child B rolls the dice and takes the corresponding number of pennies from Child A. Child A now has 13p take away the number rolled by Child B. This process continues round the circle. Repeat until everyone has had three rolls of the dice. At the end of the game, ask each child to count the number of coins they have on their plate. The child with the most coins is the winner. More able children can be encouraged to exchange their pennies for coins of a greater denomination, for example, 5 × 1p for a 5p. Ask the children **Who do you think has most coins / fewest coins? Do you think that anyone has the same number of coins? Count the number of coins you have on your plate.**

TASK 7

Speed building

- Provide the children with a collection of 3D objects – wooden blocks used in loose parts play are ideal. Encourage the children to explore the different types of blocks in a play scenario and provide lots of opportunities for handling the objects informally before embarking on this assessment.

- Ask **What kind of blocks can you see? Do you know what they are called**? **Can you sort these blocks? How did you sort them?** Check whether the children can explain the criteria they have used. Now ask **Can you sort the blocks in a different way?**

- The children may sort by the actual physical nature of the object (for example, all the cuboids together, cubes together) or they might sort by the properties of the object; for example, whether the objects roll or not. The way the children approach the task will depend on previous experiences and also on the foci used when teaching shape and sorting.

- Some questions to explore. **Which 3D object can be used to build the tallest, most stable tower? How many bricks tall is the tower? What objects did you use?**

Related task

- Children are challenged to build the most stable tower they can, using the (most appropriate) blocks provided, applying their acquired knowledge of 3D objects.

- Ask **Can you estimate how tall a tower of blocks you can build in 10 seconds, 30 seconds or 1 minute?** Encourage the children to record their prediction on a white board.

- Set a sand timer and ask the children to commence building. How close in number, to their prediction, was their actual tower?

- **Can you tell me the best objects to build with?**

End of early level assessment 4

One caterpillar and two caterpillars

Context

Most children will have experienced Eric Carle's wonderful story of *The Very Hungry Caterpillar*. It can, however, be re-read in preparation for this assessment if required. It makes useful links to Science through its illustration of the metamorphosis of a caterpillar into a butterfly.

Resources
Resources provided online:
Day and night images • Food images • Outline of a symmetrical butterfly shape
Resources needed and available in the setting:
The Very Hungry Caterpillar by Eric Carle • Counters • Cardboard/paper circles • Strips of card • Pens • Paper

Tasks	Es & Os	Benchmarks
Task 1 **a** Growing caterpillars	MTH 0-13a I have spotted and explored patterns in my own world and the wider environment and can copy and continue these and create my own patterns.	• Copies, continues and creates simple patterns involving objects, shapes and numbers.
b Sorting caterpillars	MNU 0-11a I have experimented with everyday items as units of measure to investigate and compare sizes and amounts in my environment, sharing my findings with others.	• Compares and describes lengths, heights, mass and capacities using everyday language including longer, shorter, taller, heavier, lighter, more and less.
c Measuring caterpillars	MNU 0-02a I have explored numbers, understand that they represent quantities, and I can use them to count, create sequences and describe order.	• Uses one-to one correspondence to count a given number of objects to 20. • When counting objects, understands that the number name of the last object counted is the name given to the total number of objects in the group.
d Listen and respond	MTH 0-16a I enjoy investigating objects and shapes and can sort, describe and be creative with them.	• Recognises, describes and sorts common 2D shapes and 3D objects according to various criteria, for example, straight, round, flat and curved.
Task 2 Time	MNU 0-02a I have explored numbers, understanding that they represent quantities, and I can use them to count, create sequences and describe order.	• Doubles numbers to a total of ten mentally.
	MNU 0-10a I am aware of how routines and events in my world link with times and seasons, and have explored ways to record and display these using clocks, calendars and other methods.	• Names the days of the week in sequence, knows the months of the year and talks about features of the four seasons in relevant contexts. • Links daily routines and personal events to time sequences.

End of early level assessment 4

Tasks	Es & Os	Benchmarks
	MNU 0-20c I can use the signs and charts around me for information, helping me plan and make choices and decisions in my daily life.	• Interprets simple graphs, charts and signs and demonstrates how they support planning, choices and decision-making. • Applies counting skills to ask and answer questions and make relevant choices and decisions based on the data.
Task 3 Butterflies	MNU 0-02a I have explored numbers, understanding that they represent quantities, and I can use them to count, create sequences and describe order.	• Orders all numbers forwards and backwards within the range of 0–20. • Uses the language of before, after and in-between.
	MTH 0-17a In movement, games and technology I can use simple directions and describe positions.	• Understands and correctly uses the language of position and direction, including in front behind, above below, left right, forwards and backwards to solve simple problems in movement games.
	MTH 0-19a I have had fun creating a range of symmetrical pictures and patterns using a range of media.	• Identifies, describes and creates symmetrical pictures with one line of symmetry.

TASK 1

a) Growing caterpillars

• Provide the children with a paper or card circle (to represent a caterpillar face) and a collection of counters; ensure that there are three or four choices of colours, in reasonable quantities. Have a caterpillar already prepared for the children to copy. Ask **Can you copy the pattern shown on my caterpillar?** Encourage the children to use the counters to make their own caterpillar. Ask **What colour of counter will you add next to make the caterpillar longer? What will the colour of the next two counters after that be?** Encourage the children to show you their thinking by continuing the pattern.

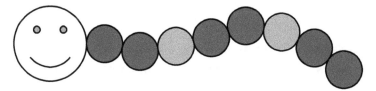

• Ask **Can you use counters to make a different caterpillar with a repeating pattern?** Provide the children with a cardboard face as before. Encourage the children to explore several possible patterns and prompt them, if necessary, to use three or four colours after their first attempt, for example:

red   green   blue   red   green   blue, etc.
green   green   red   blue, etc.
or red   green   blue   blue, etc.

Ask them to describe their pattern.

b) Sorting caterpillars

- Provide six strips of fine card or paper cut to different lengths. Ask **Can you sort these 'caterpillars' in order of length (shortest to longest or longest to shortest)?**

  Look for children who realise that they must have all the caterpillar strips lining up from a fixed point so that they can be compared fairly. Also look for children who may opt to use cubes or paper clips to measure each caterpillar.

c) Measuring caterpillars

- Provide the children with six caterpillars created from varying numbers of same-sized segments. The idea is for the children to use the number of segments to determine the 'measurement' rather than comparing their lengths by measuring them. Ensure that not all caterpillars are 'facing' the same direction. **Which caterpillar is the longest/shortest? Can you arrange the caterpillars in order from longest to shortest/shortest to longest?** Ask questions which challenge the children to compare the caterpillars; for example, **Which caterpillars are shorter than the grumpy caterpillar? How many caterpillars are longer than the smiling caterpillar?** etc.

- **What can you tell me about these different caterpillars?** The children should use their own words. Then say **Think about length, segments, and so on.**

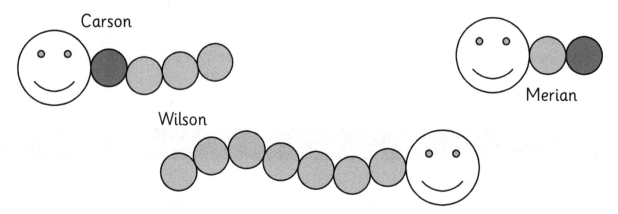

- Children may note in the relationship between caterpillars that Merian has half the number of segments that Carson has.

d) Listen and respond (this is a challenge task – beyond Early Level)

- Provide the children with a piece of paper and pencil/felt tip pen. The children must listen carefully to the instructions and draw what is described to them. Ensure that the children are familiar with any vocabulary that you might use; for example, segment, antennae. Guidance will need to be given on where to start on the page. A useful tip is to dot the paper where you want the children to begin.

  Instructions
  1. **We are going to draw a caterpillar's head. Start by drawing a circle.**
  2. **Inside the circle give the caterpillar two little circle eyes, a triangle nose and a rectangle mouth.**
  3. **On top of the caterpillar's head draw two rectangular antennae of the same size.**
  4. **Give your caterpillar seven oval segments to make its body.**
  5. **Give each segment two little legs.**

End of early level assessment 4

TASK 2

Time

- Read the story *The Very Hungry Caterpillar* again if necessary. The story covers aspects of time – days of the week and day and night. It is important that the children can access the story to recap some of the information.

- Ask **How many days of the week are there? Can you name them?**

- Ask the children further questions about the caterpillar's eating habits.

- Can the children tell you about the events in the story? **What did the caterpillar eat on Wednesday? What day comes before Sunday? On which day did the caterpillar eat some oranges? How many did he eat? How many pieces of fruit did he eat altogether on Monday, Tuesday and Wednesday? How many things did he eat on Saturday? What day comes after Thursday?** Let's work together and find out how many things the caterpillar ate by Saturday night. **Can you use marks to keep track of the items?**

- Say **Look at this table. It shows what one caterpillar ate each day.** Ask **Can you work out how many pieces of fruit two caterpillars would eat?**

	Monday	Tuesday	Wednesday	Thursday	Friday
one caterpillar	🍎	🍐🍐	🍑🍑🍑	🍓🍓🍓🍓	⬤⬤⬤⬤⬤
two caterpillars					

- Provide the children with a series of images/graphics of day and night activities (see Resource 32). Ask the children to **Sort the pictures into day and night sets.** Ask the children to discuss their choices.

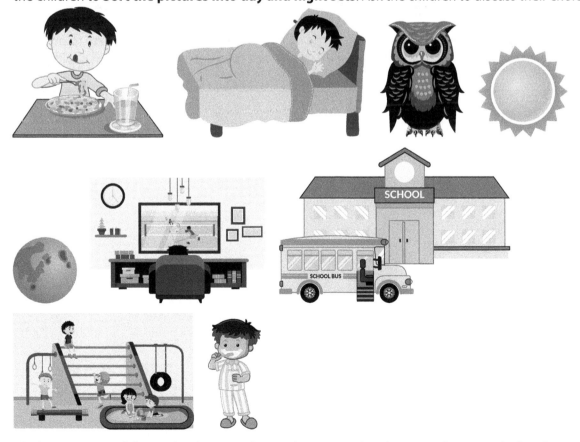

Including pictures of things that happen during daytime and night-time – for example, brushing teeth – can provoke valuable discussion.

- Remind the children about the stomachache that the caterpillar had on Saturday night. Ask **Which food choices should the caterpillar make?**

- Show the images of different types of food to the children (see Resource 33). Ask them to put a tick or a cross on their personal white board to illustrate a good food choice or a bad food choice for a caterpillar.

TASK 3

Butterflies

- Provide the children with a symmetrical outline of a butterfly (see Resource 34). Ask **Can you make a beautiful butterfly with symmetrical patterns on its wings? You may use shapes and colours in its design**. Once the children have completed their picture, ask **Can you describe the position of the shapes and colours? What shapes have you used? What colours have you used?** Look carefully at the designs made by the children. Ask them to describe the position of some of the colours/shapes that they have used in their design. For example, **What shape is next to the red triangle? What colour is above the blue square? What shape have you used at the bottom of the butterfly's wing?**

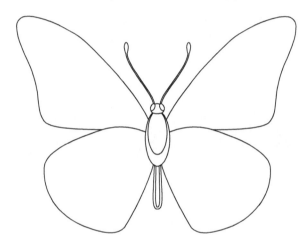

Outdoor assessment activity

- Chalk some flower heads on an area of the playground similar to the one shown here.

- Write a number inside each flower head. They should not be drawn in order. Provide the children with a starting point and tell them they are butterflies who need to visit the flowers to get nectar. They must visit the flowers in sequence; tell them which number flower to start on and which number flower to finish on. Vary the starting number and include examples which require the children to count backwards. Add challenge by getting the children to count in tens. Ask the children **What number comes after / before the flower you are on? What flower would be between ... and ... ?**

End of early level assessment 5

Using the outdoors

Context

The outdoor environment provides a space to assess children in a more relaxed and natural way. It provides a range of natural ways to link different concepts which can help children to be more forthcoming with ideas and suggestions.

Resources

Resources needed and available in the setting:

Syringes • Playground chalk • Selection of materials suitable for measuring length • Sticks • Material to screen/hide the sticks • Selection of coins • Bucket • Collection of empty containers • Lollipop sticks • Plastic bottle tops • Pebbles

---

Tasks	Es & Os	Benchmarks
Task 1 Syringe shoot	MNU-0-11a I have experimented with everyday items as units of measure to investigate and compare sizes and amounts in my environment, sharing my findings with others.	• Shares relevant experiences in which measurements of lengths, heights, mass and capacities are used, for example, in baking. • Compares and describes lengths, heights, mass and capacities using everyday language, including longer, shorter, heavier, lighter, more and less. • Estimates, then measures the length, height, mass and capacity of familiar objects using a range of appropriate non-standard units.
Task 2 Stick sort	MNU 0-01a I am developing a sense of size and amount by observing, exploring, using and communicating with others about things in the world around me. MNU 0-02a I have explored numbers, understanding that they represent quantities, and I can use them to count, create sequences and describe order. MNU 0-11a I have experimented with everyday items as units of measure to investigate and compare sizes and amounts in my environment, sharing my findings with others.	• Demonstrates skills of estimation in the contexts of number and measure using relevant vocabulary, including less than, longer than, more than and the same. • Uses ordinal numbers in real life contexts, for example, 'I am third in line'. • Compares and describes lengths, heights, mass and capacities using everyday language, including longer, shorter, taller, heavier, lighter, more and less.

End of early level assessment 5

Tasks	Es & Os	Benchmarks
Task 3 Coin drop	MNU 0-01a I am developing a sense of size and amount by observing, exploring, using and communicating with others about things in the world around me. MNU 0-11a I have experimented with everyday items as units of measure to investigate and compare sizes and amounts in my environment, sharing my findings with others. MNU 0-20a I can collect objects and ask questions to gather information, organising and displaying my findings in different ways.	• Checks estimates by counting. • Demonstrates skills of estimation in the contexts of number and measure using relevant vocabulary, including less than, longer than, more than and the same. • Compares and describes lengths, heights, mass and capacities using everyday language, including longer, shorter, taller, heavier, lighter, more and less. • Applies counting skills to ask and answer questions and make relevant choices and decisions based on the data.
Task 4 Shapes and sticks	MTH 0-16a I enjoy investigating objects and shapes and can sort, describe and be creative with them.	• Recognises, describes and sorts common 2D shapes and 3D objects according to various criteria, for example, straight, round, flat and curved.
Task 5 Pebble dominoes	MNU 0-02a I have explored numbers, understanding that they represent quantities, and I can use them to count, create sequences and describe order. MNU 0-03a I can use practical materials and can 'count on and back' to help me understand addition and subtraction, recording my ideas and solutions in different ways.	• Counts on and back in ones to add and subtract. • Doubles numbers to a total of 10 mentally. • Identifies how many in regular dot patterns, for example, arrays, five-frames, ten-frames, dice and irregular dice patterns, without having to count (subitising). • Group items recognising that the appearance of the group has no effect on the overall total (conservation of number). • Uses appropriately the mathematical symbols +, – and =. • When counting objects, understands that the number name of the last object counted is the name given to the total number of objects in the group. • Solves simple missing number problems.

End of early level assessment 5

Syringe shoot

- Syringe shoot involves the children firing water at a target using a syringe (most schools have science equipment which should have a few syringes in the pack). To allow the children to participate fully in this activity, provide some free exploration with syringes in advance of the assessment; for example, through opportunities to expel water from them, ideally outdoors! Ask the children to describe what happens when they fill the syringe and push the plunger in. Listen to the children talking as they fill the syringe. Is there evidence of mathematical vocabulary being used appropriately and in context? This exploration phase is important in helping children visualise concepts and should not be rushed. **What do you need to do to fill the syringe? What happens to the water? What makes the water go further?** This activity should provide a frame of reference for the distances that the water may travel and provide the children with some understanding of the distances to be measured.

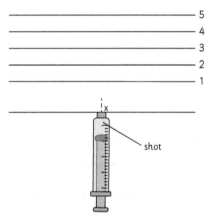

- Before commencing the activity, it is pertinent to work with the children to prepare the target zone by measuring the distance from X to 1, X to 2, etc. (see sketch) so that when they actually start the activity, they just need to record the distances. Provide a tray of items which *may* or *may not* be useful or appropriate for measuring distance. Encourage the children to sort the items into useful and not useful and then to use the useful ones to measure the distance. Can the children come up with some ideas for different materials/objects they could use to measure the distances and then carry out the measurements?

- This is an opportunity to assess mathematical vocabulary and the child's ability to use it correctly. Non-standard units of measurement to place in the tray for discussion could include – metre sticks (not to refer to measuring in metres but as 'long sticks' of equal length), bead strings (not for counting the beads but as a standard length to lay out), shapes, string, skipping ropes, lollipop sticks, shells and paces.

- The target should be a series of lines chalked on the playground. Ask one child at a time to fill a syringe and stand on the X. The child should press in the plunger as hard as they can and see how far they can shoot the water. The distance achieved is the furthermost point that the water travels. The children use the chalked score grid to allocate scores. The results could be recorded using chalk marks or numerals on the playground.

- Children may also explore this activity in other ways. A child stands on an X marked on the playground and shoots the water out of their syringe. The children can estimate how far they think they might squirt the water and then measure to confirm their prediction. This time there is no grid to aim at. An adult or child can be directed to spot where the water lands and draw a ring around the 'splat' of water. The rest of the children in the group take turns to fire the syringe/s and record the distance reached. A group discussion may ensue about who shot the water the furthest, the shortest distance, etc.

- There are lots of variations and investigations that may arise from this assessment. **Is there an optimum amount of water in the syringe? Is a half-full syringe better than a full one? Does the size of syringe matter? How can we ensure a fair test?** This assessment presents lots of opportunities for the children to demonstrate previously acquired language and problem-solving skills in a totally different context.

TASK 2

Stick sort

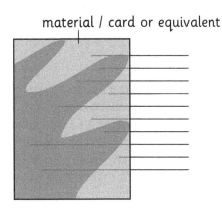

material / card or equivalent

- Prior to this assessment, collect a selection of sticks of varying lengths. Prepare the activity in advance. Lay all the sticks to be used in a row, as in the picture. Partially cover all the sticks as shown – the visible part of each stick should be arranged so that they look as though they are all the same length.

- Children take turns to remove one stick each from under the cover. This should be done carefully to prevent the cover moving and revealing what is underneath. Once each child in the group has chosen a stick from under the screen, say **Can you compare stick lengths and arrange them in order of size**? Longest to shortest or shortest to longest is acceptable. Check to see if the children realise that the sticks should all be measured from a common 'line' so that the comparison is valid. **How can we compare the sticks? Is that a fair test? If not, how can we make it fair?** If they 'stand' their sticks on the ground they can carry out a good comparison. Can the children describe where their stick is in the seriation line? For example, mine is second in the line. My stick is fifth. My stick is the same length as … My stick is next to the longest one.

End of early level assessment 5

TASK 3

Coin drop

This assessment involves a full bucket of water.

a) Explore filling a bucket with water

- Note the child's ability to use and understand appropriate mathematical vocabulary; for example, full, not full, half full, almost empty, empty, more, less, most, least, same and different. Provide opportunities for the children to work collaboratively with others. Observe what each child does and engage them in conversations about the filling process. **How full is the bucket? How many cupfuls of water do you think the bucket will hold?** Use a range of vessels, for example, a plastic cup, a plastic bottle and jugs of varying size. Provide containers that have the same capacity but are different shapes. **Why did you choose that container? Do you think that they can all hold the same amount of water? Which jug do you think will hold more than the bottle?** Allow the children time to predict how many cupfuls/jugfuls the bucket will hold. Encourage the children to count the number of cupfuls/jugfuls, etc., required to fill the bucket; this can be repeated using different containers to check if the containers contain more, less or the same. **How many cupfuls did the bucket hold? How good was your estimate? Did the bucket hold more cupfuls or less cupfuls than you thought it would? Which container do you think holds most water? How can we find out?** *In order not to waste water, rainwater collected in a water butt may be used.*

b) Coin drop challenge

- Place a coin in the centre of the bucket's base (a £2 coin is a useful target due to its size). Fill the bucket with water (see activity a). Children can take turns to drop a 1p coin into the bucket. Their aim is to get their coin on top of, or as near to, the submerged £2 coin. Encourage predictions as to how many coins they think will sit on top of the submerged coin or at least touch it. Children should then be given opportunities to test their predictions. How many children managed to predict correctly? Keep track of the number of hits using chalk marks. Does it matter which coin is used? Try using different coins. **Which coin gives the best results?** Can the children offer suggestions, for example 5p versus 1p? **Does it make a difference if you drop the coin from a great height or gently drop it in?** Observe how the children tackle the problem, listen for effective vocabulary and watch for appropriate recording.

TASK 4

Shapes and sticks

- Provide the children with several lollipop sticks and have some playground chalk to hand. Encourage the children to explore what shapes they can make using the sticks. Ask them to annotate their shape by chalking numbers they might use to describe it; for example, alongside a triangle they might mark the number 3. If this activity is taking place on grass/artificial grass then provide a selection of laminated/ wooden/plastic numbers to record their answer instead of writing the numbers down. Ask **What does the three represent?** (three sides or three corners, etc.). **Do all triangles use the same number of sticks?** Some children are aware that not all triangles look the same but it is not expected that they should name the different types of triangles (isosceles, scalene and right-angled).The children may choose to make regular and irregular 2D shapes or they may remember previous work on money/coins and, for example, attempt to make a heptagon to represent a 20p or 50p coin.

Challenge

- **Can you make your shape bigger? Can you count the number of sticks you used? Is there a pattern developing when the shape is made bigger?** This is a very open-ended task and the 'results' will depend greatly on each child's existing knowledge of 2D shapes and their ability to explore independently.

End of early level assessment 5

TASK 5

Pebble dominoes

- This assessment, once demonstrated, can become a good paired outdoor activity. Children should be encouraged to take turns at leading, questioning and providing the solution. In this way it is easy to observe the children in a more natural setting working with their peers.

- The teacher or adult observing the activity should draw several blank domino shapes on the playground and then chalk a number beside each domino. This will be the total number that the children together will be trying to make.
- Encourage the children to collect pebbles/small stones from the playground or provide a tub containing decorative pebbles.
- Set the scene for the children. It may be necessary to demonstrate this a couple of times to the children.

Game 1

- Explain to the children that this is a two-person activity. Child 1 will put a number of pebbles in the first section of the domino. Child 2 will have to put the correct number of pebbles into the second section of the domino that, when added to Child 1's pebbles, makes the total chalked on the playground next to the domino. Together they should check the answer. Ensure that the children take turns being first to place the pebbles into the domino. Provide challenge by asking the children to represent the numeral total as a series of tally marks instead of writing it.

- Ask **How did you make your target number? Is there a different way to make that number? Can you write number sentences to match what you did?**

Things to look for include:

- o children working together to solve the task
- o evidence of children having to touch/count the pebbles rather than being able to subitise
- o children counting from one rather than counting on
- o children recognising the question as a known fact
- o children understanding that the first addend can be placed in either section of the domino without affecting the total.

Game 2

- This is an extension of Game 1, also played in pairs. Child 1 chooses both the total to be made and the number of pebbles to go into one section of the domino. Child 2 must place the correct number of pebbles in the empty section of the domino to make the total. For example, Child 1 decides that the total will be 9 and writes 9 next to the domino. Child 1 then puts 4 pebbles in one section of the domino. Child 2 decides how many pebbles need to be added to 4 to make 9 and places that amount in the other section. The children should record the corresponding number sentence underneath the domino. The children should take turns at going first and second.

End of early level assessment 6

Goldilocks' big bake off challenge

Context

Goldilocks wants to bake some biscuits for the three bears – Mummy Bear, Daddy Bear and Baby Bear (see Resource 35). She has a recipe for biscuits; however, the recipe only makes the right number of biscuits that Mummy Bear likes to eat – too many for Baby Bear and too few for Daddy Bear.

Perhaps you can help Goldilocks with her baking challenge and find out how she could make the correct amount of biscuits for Baby Bear and Daddy Bear?

Resources

Resources provided online:

Pictures of the three bears representing their size differences • Copy of the recipe • Table showing the number of biscuits eaten • Ingredients table • Biscuit box image •   Biscuit table • Round paper circles to represent three different-sized biscuits •Large paper circle to represent a plate

Resources needed and available in the setting:

Traditional tale of Goldilocks and the three bears • Selection of concrete materials

---

Task	Es & Os	Benchmarks
Task 1 Recipe for success	MNU 0-02a I have explored numbers, understanding they represent quantities, and I can use them to count, create sequences and describe order. MNU 0-10a I am aware of how routines and events in my world link with times and seasons, and have explored ways to record and display these using clocks, calendars and other methods.	• Identifies and recognises numbers from 0 to 20. • Uses ordinal numbers in real life contexts, for example 'I am third in line'. • Recognises, talks about and where appropriate, engages with everyday devices used to measure or display time, including clocks, calendars, sand timers and visual timetables.
Task 2 How much does Goldilocks need?	MNU 0-03a I can use practical materials and 'count on and back' to help me understand addition and subtraction, record my ideas and solutions in different ways.	• Doubles numbers to a total of 10 mentally • Partitions quantities to 10 into two or more parts and recognises that this doesn't affect the total • Shares out a group of items equally into smaller groups. • Uses appropriate vocabulary to describe halves.
Task 3 How much will we need?	MNU 0-07a I can share out a group of items by making smaller groups and can split a whole object into smaller parts.	

Task	Es & Os	Benchmarks
Task 4 Decorating the biscuits	MNU 0-02a I have explored numbers, understanding that they represent quantities, and I can use them to count, create sequences and describe order. MNU 0-03a I can use practical materials and can 'count on and back' to help me understand addition and subtraction recording my own ideas and solutions in different ways. MTH 0-13a I have spotted and explored patterns in my own and the wider environment and can copy and continue these and create my own patterns.	• Uses one-to-one correspondence to count a given number of objects. • Recalls the number sequence forwards within the range 0-30, from any given number. • Doubles numbers to a total of ten mentally. • Adds and subtracts mentally to 10. • Copies, continues and creates simple patterns involving objects, shapes and numbers. • Explores recognises and continues simple number patterns.
Task 5 Hungry bears a), b) and c)	MNU 0-01a I am developing a sense of size and amount by observing, exploring, using and communicating with others about things in the world around me. MNU 0-03a I have explored numbers, understanding they represent quantities, and I can use them to count, create sequences and describe order. MTH 0- 13a I have spotted and explored patterns in my own and the wider environment and can copy and continue these patterns and create my own patterns. MNU 0-20a I can collect objects and ask questions to gather information, organising and displaying my findings in different ways.	• Recognises the number of objects in a group, without counting (subitising) and uses this information to estimate the number of objects in other groups. • Checks estimates by counting. • Demonstrates skills of estimation in the contexts of number and measure using relevant vocabulary, including less than, longer than, more than and the same. • Identifies 'how many' in regular dot patterns, for example, arrays, five-frames, ten-frames, dice and irregular dot patterns, without having to count (subitising). • Groups items recognising that the appearance of the group has no effect on the overall total. • Solves simple number problems. • Copies, continues and creates simple patterns involving objects, shapes and numbers. • Contributes to concrete or pictorial displays where one object or drawing represents information. • Applies counting skills to ask and answer questions and make relevant choices and decisions based on the data.

End of early level assessment 6

TASK 1

Recipe for success

- Show/read the recipe for Mummy Bear's biscuits (see Resource 36) to the children. This recipe makes 10 biscuits, which is just right for Mummy Bear to eat. This is a fictitious recipe made to suit the task!

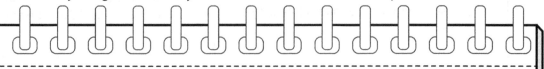

Mummy Bear's biscuits

Suitable for one mummy bear (10 biscuits)

Ingredients

4 tablespoons of sugar

10 tablespoons of white flour

6 eggs

2 tablespoons of butter

Method

1. Place the sugar and butter into a mixing bowl and beat them together.
2. Add the eggs and flour. Mix into a dough.
3. Use this dough ball to make 10 small balls.
4. Place the small dough balls onto a baking tray.
5. Place the tray in a medium temperature oven at Gas Mark 4 for 30 minutes.

A discussion can be had about the temperature of the oven but using Gas Mark terminology avoids handling larger numbers. Gas Mark 3/4 is equivalent to 160–190 degrees Celsius or 325–375 degrees Fahrenheit.

- Ask the children **Which ingredient was first to go into the bowl? Which was second? Which was last? Which ingredient did we use most tablespoons of?**
- Ask the children **What can we use to measure the time the biscuits are in the oven?**

TASK 2

How much will Goldilocks need?

- Goldilocks also wants to bake biscuits for Daddy Bear and Baby Bear. She wants to make the correct number for each of them.
- Look at Resource 37. Goldilocks knows that Daddy Bear likes double the number of biscuits that Mummy Bear likes and Baby Bear likes half the number of biscuits that Mummy Bear likes. Ask the children to **Work out how many biscuits would be just right for Daddy Bear to eat. How many biscuits would be just right for Baby Bear to eat?**

| | 10 biscuits | ? |

- Provide concrete materials including counters, beadstrings, Rekenrek and cubes to enable the children to explore doubling. Also, provide paper and pens/pencils. Observe how they solved the task. Ask the children if they used known facts. Did they act out/share the materials or take more materials? Did any child record using pictures? Ask the children to talk about what they did and to justify their thinking.
- Ask the children to think about **What number of biscuits would be needed if two Mummy Bears were coming to tea? Three Baby Bears?**

TASK 3

How much will we need?

- The table shows how much of each ingredient is needed to make Mummy Bear's biscuits (see also Resource 38).

Ingredient	Mummy Bear	Daddy Bear	Baby Bear
sugar	4		2
flour	10		
eggs	6	12	
butter	2		

We know that Daddy Bear needs double the amount of Mummy Bear's ingredients and Baby Bear needs half the amount of Mummy Bear's ingredients. **Help me fill in the numbers in the table.**

- Provide concrete materials including counters, beadstrings, Rekenrek, cubes to enable the children to explore doubling. Also, provide paper and pens/pencils. Observe how they solved the task. Ask the children if they used known facts. Did they act out/share the materials or take more materials? Did any child use pictures for recording? Ask the children to talk about what they did and to justify their thinking.

TASK 4

Decorating the biscuits

- Mummy Bear likes five raisins on each of her biscuits. **How many raisins would be needed for two biscuits? What about three biscuits? What about five biscuits? Can you count out five objects? 10 objects? Can you show this using counters (and/or) five-frames or ten-frames?**

- Baby Bear likes two cherries on his biscuits. **How many cherries will be needed for two biscuits? What about three biscuits? What about five biscuits? Can you explain to me your thinking for how you solved this problem?**

- Daddy Bear likes little stars on his biscuits. He likes to have 10 stars on each biscuit. **How many stars would you need to put on three biscuits?**

- **Can you show this using practical materials?** Observe how the child counts the stars, for example, in ones or in tens.

TASK 5

Hungry bears

a) Show the children the picture of the different-sized biscuits in a box for 10 seconds (also see Resource 39). Ask them **How many biscuits can you see altogether?** Explain that Daddy Bear has the biggest biscuits, Mummy Bear the middle-sized biscuits and Baby Bear the small biscuits. **Does Daddy Bear have more biscuits than Mummy Bear? Who has the most biscuits?**

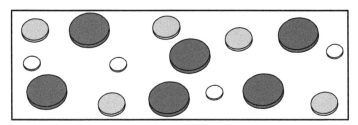

○ Baby Bear's biscuits

○ Mummy Bear's biscuits

● Daddy Bear's biscuits

b) The pictures below show us the number of biscuits that each bear ate. Use the information provided to complete a pictogram (see Resource 40), illustrating the number of biscuits eaten.

20		⬭
19		⬭
18		⬭
17		⬭
16		⬭
15		⬭
14		⬭
13		⬭
12		⬭
11		⬭
10		⬭
9		⬭
8		⬭
7		⬭
6		⬭
5		⬭
4		⬭
3		⬭
2		⬭
1		⬭
Baby Bear	Mummy Bear	Daddy Bear

- Ask the children to **Count how many there are of each size (1:1). Now record the totals. Can you calculate how many biscuits Baby Bear has eaten**? Remind them what number of biscuits each bear started off with. (Baby Bear started with 5, ate 4; Mummy Bear started with 10, ate 5; Daddy Bear started with 20, ate 20.) Using the information displayed, discuss with the children how many biscuits were left for each bear. Daddy Bear's column could be completed, i.e. 20 biscuit shapes above Daddy Bear's name. Ask **How many biscuits does Daddy Bear have now?** Provide counters or cookie-type circles to enable the children to investigate how many biscuits Mummy Bear and Baby Bear had left.

c) Provide the children with circles of three sizes representing the three different sizes of biscuits (see Resource 41) and a large paper circle to represent a plate (or see Resource 42). **Ask the children to arrange the biscuits on a plate using a repeating pattern of their choice. There must be at least one biscuit on the plate for each bear.**

End of early level assessments: marking guidance

Assessment 1: Dinosaurs

Assessment	Topic	Question
Task 1	Data and analysis MNU 0-20b	**On track** The child can identify how the dinosaurs have been sorted, use their knowledge of colour, size and other attributes to sort the same set of dinosaurs in a different way and justify their choices using appropriate vocabulary. **Review** The child can only sort using a given criterion or defaults to the same criterion in response to each request to sort.
Task 2	Number and number processes MNU 0-02a Properties of 2D shapes and 3D objects MTH 0-16a	**On track** The child can accurately count the number of each shape used to construct their dinosaur, recognising that the final number said is the total amount. **Review** The child is unsure of the number sequence and/or is unable to use one-to-one correspondence to count the shapes. **On track** The child can identify and name each shape used and describe them using appropriate vocabulary. For example: side, straight, round. **Review** The child needs support to identify and name all or some common 2D shapes: square, circle, triangle, rectangle.

End of early level assessments: marking guidance

Assessment	Topic	Question
Task 3	Number and number processes MNU 0-02a Measurement MNU 0-11a Data and analysis MNU 0-20a	**On track** The child can order the dinosaurs, on a 0–30 number line, to reflect the differences in their lengths. **Review** The child lacks confidence in the number word sequence and so finds it difficult to arrange the dinosaurs in order numerically. **On track** The child can, from looking at the data in the tables provided, compare and order the dinosaurs in order of length and can use appropriate vocabulary when answering questions. **Review** The child does not understand the vocabulary of measurement; for example, longer and shorter; they may only understand when the concept is described as long and not long, or short and not short. **On track** Using a selection of different-sized dinosaurs, the child can estimate then measure their heights in non-standard units. The child can then place the dinosaurs in height order. **Review** The child is unable to make realistic estimates about the relative heights of the dinosaurs. The child does not appreciate that chosen units of measurement should be the same size and/or has difficulty lining up the chosen units without overlaps or gaps. The child may be unable to order the dinosaurs according to height. **On track** The child can confidently interpret the data contained in the table, explain their thinking and demonstrate the application of their knowledge by correctly ordering the dinosaurs. **Review** The child is unable to access the data contained within the table; they may require manipulatives, such as cubes, to build the relative heights.
Task 4a and 4b	Estimation and rounding MNU 0-01a	**On track** The child can make a realistic estimate of the number of objects being displayed/flashed. **Review** The child struggles to subitise small amounts and attempts to count in ones. The child is unable to interpret questions which require them to draw comparisons, for example when asked to identify how many more/less.

End of early level assessments: marking guidance

Assessment	Topic	Question
Task 5a and 5b	Patterns and relationships MTH 0-13a	**On track** The child can copy, continue and make their own repeating AB and AAB patterns, justify their choices of materials and talk about the repeating unit of their patterns. The child can draw their own repeating patterns. The child can identify and extend simple number patterns. **Review** The child is unable to recognise the basic unit of the pattern and so is unable to make or continue repeating patterns.
Task 6	Fractions, decimal fractions and percentages MNU 0-07a	**On track** The child can share the 'ferns' fairly (pictures or concrete materials) and justify why what they have done is fair. **Review** The child finds it difficult to 'share fairly', distributing some 'ferns' to each dinosaur without apportioning them equally, and/or does not have the appropriate vocabulary to describe their actions.
Task 7	Measurement MNU 0-11a Angle, symmetry and transformation MTH 0-17a	**On track** The child can use appropriate concrete materials (for example, interlocking cubes) to investigate and compare the heights of the dinosaurs. They can then use this knowledge to build an appropriate structure for the dinosaur to fit inside. **Review** The child does not understand that, when comparing height/ length the objects must be aligned accurately, nor appreciate that the chosen non-standard units must be of equal size. The child may complete the first part of the task correctly but be unable to use what they have done to build an appropriate structure. **On track** The child can use appropriate positional language when engaging in this task. For example: in front, behind, above, below. **Review** The child is unable to use appropriate positional language without prompting or support.
Task 8	Money MNU 0-09a	**On track** The child is able to recognise coins (1p, 2p, 5p and 10p) and can apply addition and subtraction skills in relation to money; the child can pay the exact value for items up to 10p. **Review** The child is unable to identify all or some of the following coins (1p, 2p, 5p and 10p) and/or does not appreciate their value. For example, the child does not understand that one 5p coin has the same value as five 1p coins, two 2p coins plus one 1p coin, etc.

End of early level assessments: marking guidance

Assessment 2: The party

Assessment	Topic	Question
Task 1 Happy Birthday Teddy	Time MNU 0-10a Data and analysis MNU 0-20a	**On track** The child knows that there are 7 days in a week, can recite them in the correct order and is able to answer questions relating to day before/after. The child knows the months of the year and can work with others to sequence them. **Review** The child is unable to recite the days of the week in sequence. The child omits, repeats or switches some of the days so is unable to accurately recall the number of days. The child is unable to provide the day before or after a given day. **On track** The child can apply counting skills to ask and answer questions about Teddy's birthday and the birth dates of other children in the class. The child can contribute to the construction of a group/class pictogram showing the spread of group/class birthdays. **Review** The child can, with support, contribute to the pictorial display by adding their completed birthday symbol to their birthday month but has difficulty interpreting the birthday pictogram to answer questions; for example, Which month has the most birthdays?
Task 2 **a** Invitations **b** Making a clock	Time MNU 0-10a Number and number processes MNU 0-02a Properties of 2D shapes and 3D objects MTH 0-16a	**Task a** **On track** The child can represent o'clock times on analogue and digital clocks and discuss these using appropriate language. **Review** The child is not secure in recording or displaying o'clock on an analogue/digital template. The child cannot talk about the different templates. **Task b** **On track** The child can set analogue clock hands to show o'clock; long hand for minutes and short hand for the hour. **Review** The child cannot set the hands of the clock to depict o'clock times. **On track** The child can write the numbers 1–12 in sequence, on a circular display, to represent an analogue clock. **Review** The child is unable to write all/any of the numbers 1–12 in sequence; if given pre-prepared numbers the child cannot order them.

End of early level assessments: marking guidance

Assessment	Topic	Question
		On track The child can name the 2D shape used to make their clock face. For example, 'I have used a circle to make my clock face.' The child can use appropriate mathematical language to make and describe the clock hands: long, short, minute hand, hour hand. **Review** The child can make/choose a clock face but cannot name the 2D shape it represents and/or create and describe its features.
Task 3 Planning a party	Number and number processes MNU 0-02a Fractions, decimal fractions and percentages MNU 0-10a	**On track** The child can represent the problem with concrete materials or by drawing. The child understands the concept of 'fair share' and can share the party items equally. The child spontaneously counts the items to answer questions such as, 'How many more?' and 'Are there enough?' **Review** The child may 'share' the party items among Teddy and his friends but does not appreciate the need for fair shares, that is, the need to give each 'person' an equal amount. The child may be able to count the items but is unable to describe their actions and/or answer questions such as, 'How many more?' and 'Are there enough?'
Task 4 Birthday money	Number and number processes MNU 0-03a Money MNU 0-09a	**On track** The child recognises the coins listed and can show different ways of representing the cost of the toys illustrated. The child can represent and solve money-related addition and subtraction problems. **Review** The child may be able to recognise some, but not all, of the coins listed. The child does not appreciate that the same amount can be represented in different ways and so is unable to solve money-related word problems within 10p.

Assessment 3: Playing games

Assessment	Topic	Question
Task 1 Board games	Estimation and rounding MNU 0-01a Number and number processes MNU 0-02a Number and number processes MNU 0-03a	**On track** The child can recognise the number of dots on the dice (subitise) without counting. The child can use the terms more/less than to describe the number of dots. **Review** The child requires to touch/count the spots on the dice and/or cannot identify which has more/less. **On track** The child can recognise the written number sequence 1–30 and can use one-to-one correspondence to keep track of each character's score on a numbered track. **Review** The child is unable to correspond each dot on the dice with a place on the number track and/or is unable to read the numerals 17 and 18 correctly. **On track** The child can accurately count on and back in ones on a numbered track. **Review** The child uses the number track incorrectly, counting numerals rather than jumps.
Task 2 Sudoku	Patterns and relationships MTH 0-13a	**On track** The child can identify the pattern required for the Sudoku grid and describe how they completed it. **Review** The child may be able to copy a given Sudoku pattern but is unable to complete a partially filled or empty grid.

End of early level assessments: marking guidance

Assessment	Topic	Question
Task 3 Magnetic fishing	Numbers and number processes MNU 0-03a Data and analysis MNU 0-20a	**On track** The child can calculate the total score (for each child) – either mentally or using concrete materials. **Review** The child is unable to recall addition bonds to 10 and/or the child is unable to lay out concrete materials in an ordered fashion and so makes errors within the total count. **On track** The child can use their knowledge of quantity to answer questions based on the information in the table. They understand the terms highest/lowest/same, in relation to the scores. **Review** The child is unable to interpret the table and/or does not understand the terms highest/lowest/same in relation to the scores. **On track** The child can choose an appropriate way to record his or her own score and compare it with other children's scores. The child can judge which score is more/less. **Review** The child can use concrete materials or make marks to record his/her own score but cannot compare it with other's scores.
Task 4 Wet sponge splosh	Data and analysis MNU 0-20a	**On track** The child can interpret the data displays and answer simple questions about them. The child can compare two quantities, either using concrete materials or pictorially, to work out more/less/difference. The child can represent and record their own data pictorially. **Review** The child does not understand the language highest/lowest/difference between and/or is unable to identify a strategy for comparing two quantities.
Task 5 Hungry mouse	Angle, symmetry and transformation MTH 0-17a	**On track** The child can accurately count the number of steps taken and use appropriate positional language – forward/backward/left/right – to describe the direction. **Review** The child can count the number of steps but confuses left/right and/or forwards/backwards.

End of early level assessments: marking guidance

Assessment	Topic	Question
Task 6 Taking coins	Estimation and rounding MNU 0-01a Number and number processes MNU 0-02a	**On track** The child can estimate who has more/less coins and check the accuracy of their predictions by counting them. **Review** The child is unable to describe or compare the collections of coins. **On track** The child can subitise the dot patterns on the dice. **Review** The child touch-counts the dot patterns on the dice. **On track** The child can apply addition and subtraction skills using 1p coins. **Review** The child is unable to count the coins accurately using one-to-one correspondence and/or may miss out numbers in the sequence or confuse the number order.
Task 7 Speed building	Estimation and rounding MNU 0-01a Properties of 2D shapes and 3D objects MTH 0-16a Data and analysis MNU 0-20b	**On track** The child can make a reasonable estimate of the size of tower they can build before the sand timer runs out, then check by counting. **Review** The child's estimate is unrealistic. They may miscount by touching some blocks more than once, or by reciting the number names in the wrong order. **On track** The child can name and describe the 3D objects provided and is able to apply their knowledge of 3D objects to build a stable tower. The child can talk about the features of the shapes they have used using appropriate vocabulary, for example, straight, round, flat and curved. **Review** The child may be able to build a stable tower but is unable to name the 3D objects used and/or describe their key features. **On track** The child can sort the blocks using both given and their own criteria and justify their choices using appropriate vocabulary (colour/size/shape). **Review** The child can only sort using a given criteria; they cannot justify their choice.

Assessment 4: One caterpillar and two caterpillars

Assessment	Topic	Question
Task 1 **a** Growing caterpillars	Pattern and relationships MNU 0-13a	**Task a** **On track** The child can copy and continue a simple repeating pattern; they are able to design and describe their own repeating pattern. **Review** The child can copy a pattern but may be unable to continue a pattern or create their own pattern.
b Sorting caterpillars	Measurement MNU 0-11a	**Tasks b and c** **On track**
c Measuring caterpillars	Numbers and number processes MNU 0-02a	The child accurately compares and orders the 'caterpillars' by length and can use appropriate vocabulary to describe them, that is, longer/shorter/longest/shortest/before/after/in-between (Task b). The child appreciates that same-sized units can be used to find out which is longer/shorter, without the need to match and compare (Task c).
d Listen and respond	Properties of 2D shapes and 3D objects MTH 0-16a	**Review** The child does not appreciate that to make a fair comparison the caterpillars must be aligned with the same fixed point (Task b). The child recognises that the lengths are different but is unable to explain why. The child does not trust counting in non-standard units as a reliable way to compare length instead preferring to cut out and 'match' the caterpillars (Task c). **Task c** **On track** The child can apply their knowledge of number (doubles, conservation of number, number order) to work out the difference between the caterpillar lengths. The child can subitise the number of segments but may count to check. **Review** The child counts the segments; the child knows some numbers but may not put them in order; they may allocate zero to the first segment counted. **Challenge (beyond Early Level) (task d)** The child can follow the instructions given to draw recognisable representations of the named shapes, that is, circle, triangle, rectangle, oval.

End of early level assessments: marking guidance

Assessment	Topic	Question
Task 2 Time	Time MNU 0-10a Data and analysis MNU 0-20c	**On track** The child can name the days of the week in order and knows how many days there are in a week. **Review** When asked to name the days of the week in sequence, the child omits, repeats or 'switches' some days and so is unable to accurately recall the number of days there are in a week. **On track** The child can apply counting skills to answer questions. They can interpret a simple table and are able to make decisions and choices based on the information given. **Review** The child is unable to keep track of the number of items counted and so is unable to accurately answer questions about the data recorded.
Task 3 Butterflies	Angle, symmetry and transformation MTH 0-17a Angle, symmetry and transformation MTH 0-19a	**On track** The child can create a roughly symmetrical picture with one line of symmetry and describe it. **Review** The pattern created by the child is not a reflection, for example the pattern is rotated, enlarged or translated (slid along). **On track** The child can use the language of position and direction such as forwards/backwards/left/right/above/below to describe the pathway taken through the flowers. **Review** The child may be able to follow the number order but be unable to describe accurately the directions taken.

End of early level assessments: marking guidance

Assessment 5: Using the outdoors

Assessment	Topic	Question
Task 1 Syringe shoot	Number and number processes MNU 0-02a Measurement MNU 0-11a	**On track** The child shows accuracy in counting the number of items laid out. **Review** The child is unable to use one-to-one correspondence and/or does not appreciate that the last object counted is the name given to the total number of objects in the group. **On track** The child can suggest and use an appropriate non-standard unit for measuring distance. The child can describe distance and capacity using appropriate measurement language; for example, full, empty, more, less. **Review** Suggested unit of measurement indicates an unrealistic perception of size and distance. The child does not appreciate that chosen units should be of equal size and/or is unable to draw comparisons using appropriate vocabulary.
Task 2 Stick sort	Estimating and rounding MNU 0-01a Number and number processes MNU 0-02a Measurement MNU 0-11a	**On track** The child can work with others to compare and order the set of sticks. They openly use the language of comparison. **Review** The child may require peer support in this task and does not appear to use the language of comparison, for example, longer/shorter than. **On track** The child can use ordinal numbers to describe the position of their stick. **Review** The child can compare and order the sticks but is unable to describe their position; for example, third. **On track** The child is confident in comparing and describing lengths and can use terms such as longer and shorter. **Review** The child may be able to describe their stick as 'long' or 'short' but is unable to draw comparisons.
Task 3 Coin drop	Estimation and rounding MNU 0-01a	**On track** The child can apply his/her numeracy skills in context and make realistic estimates, justifying his/her choices and answers. **Review** The child does not give reasonable estimates or predictions.

End of early level assessments: marking guidance

Assessment	Topic	Question
	Measurement MNU 0-11a Data and analysis MNU 0-20a	**On track** The child understands and can use appropriate mathematical vocabulary (for example, full/empty, holds more/holds less) to compare the capacity of different containers, and estimate and measure using non-standard units. **Review** The child may not understand and/or be unable to use comparative mathematical language related to capacity. They may make unrealistic estimates and/or may not be concerned about whether each 'cupful' (that is, non-standard unit) is the same size. **On track** The child can tally the number of coins dropped near the target and can offer predictions related to developing the game. **Review** The child finds it hard to keep a tally of the coins near the target as touch-counting isn't an option.
Task 4 Shapes and sticks	Properties of 2D shapes and 3D objects MTH 0-16a	**Challenge** The child can create common 2D shapes (square, triangle, rectangle) and write or match an appropriate number which shows some understanding of its properties. For example, the child can recognise that, although a rectangle has four sides, they need six sticks to make it because the sticks are the same length. The child can create their own irregular 2D shapes and describe them using language such as sides and corners.
Task 5 Pebble dominoes Games 1 and 2	Numbers and number processes MNU 0-02a Numbers and number processes MNU 0-03a	**On track** The child can work collaboratively, ask questions, take turns and count accurately when completing the dominoes. They can show their knowledge of addition bonds to 10 and can record their findings accurately using the appropriate symbols. They can talk confidently about what they have been investigating. **Review** The child may rely on a peer partner to guide them through the task. The child may be unable to do one or more of the following: - subitise small quantities - count out the correct number of pebbles to make the required total - write an addition number sentence to match the completed domino. **On track** In the extension activity, the child can find the missing addend by counting on and write an addition number sentence to match the completed domino. **Review** The child does not trust 'counting on' and so counts from one to solve the problem. The child believes the missing addend can be found by adding the known addend to the total.

End of early level assessments: marking guidance

Assessment 6: Goldilocks' big bake off challenge

Assessment	Topic	Question
Task 1 Recipe for success	Numbers and number processes MNU 0-02a Time MNU 0-10a	On track – The child understands the meaning of the numbers used in the recipe and so can say which ingredient Mummy Bear used most of. The child can use the language of ordinal number (for example, first/second/last) to describe the sequence in which the ingredients went into the bowl. Review – The child knows that the quantities are not the same but cannot suggest what ingredients there were more/less of. The child cannot correctly respond to questions involving the use of ordinal numbers. On track – The child can suggest an appropriate timing device for measuring the time the biscuits are in the oven, for example, phone/clock. Review – The child is unable to suggest an appropriate timing device or makes an unrealistic suggestion.
Task 2 How much does Goldilocks need?	Numbers and number processes MNU 0-03a Fractions, decimal fractions and percentages MNU 0-07a	On track – When investigating the quantities needed, the child can explain and justify their thinking and use appropriate vocabulary to describe their actions (more for daddy bear/less for baby bear). The child understands what it means to 'double' and 'half' a quantity and can devise a strategy (for example, use practical materials, act it out, draw a picture or use known facts) to double and half the number given. Review – The child is unable to select and use an appropriate method to double and/or halve the quantities presented. The child may 'share' concrete materials according to the size of the perceived size of the characters, giving a few to baby bear, more to mummy bear and the greatest amount to daddy bear, but does not halve or double the amounts accurately.
Task 3 How much will we need?	Numbers and number processes MNU 0-03a Fractions, decimal fractions and percentages MNU 0-07a	On track – When investigating the quantities needed, the child can explain and justify their thinking and use appropriate vocabulary to describe their actions (more for daddy bear/less for baby bear). The child understands what it means to 'double' and 'half' a quantity and can devise a strategy (for example, use practical materials, act it out, draw a picture or use known facts) to double and half the number given. Review – The child is unable to select and use an appropriate method to double and/or halve the quantities presented. The child may 'share' concrete materials according to the size of the perceived size of the characters, giving a few to baby bear, more to mummy bear and the greatest amount to daddy bear, but does not halve or double the amounts accurately.

End of early level assessments: marking guidance

Assessment	Topic	Question
Task 4 Decorating the biscuits	Number and number processes MNU 0-02a MNU 0-03a Patterns and relationships MTH 0-13a	On track – The child can apply counting skills in the context of decorating the biscuits using, for example, empty five-frames and ten-frames where they can work systematically and in an ordered way grouping the objects in twos, fives and tens making for an easier count. The child can recognise and continue simple number patterns (with or without practical materials). Although this task involves counting objects up to 30, it should be noted that if the child can count up to 20 items using one-to-one correspondence, the child has achieved the expected standards according to the Early Level Benchmarks. Review – The child has difficulty counting out practical materials for Daddy Bear's and Mummy Bear's biscuits; the child may be able to correctly represent the total for one or two of Baby Bear's biscuits but not for three or five biscuits. The child does not see a number pattern emerging.
Task 5 Hungry bears a), b) and c)	Estimating and rounding MNU 0-01a Number and number processes MNU 0-02a Data and analysis MNU 0-20a Patterns and relationships MTH 0-13a	a) On track – The child can make a realistic estimate of how many biscuits are shown, by subitising. The child may identify subsets of different-sized circles and can use this to justify their answer. The child can record their answer. Review – The child does not provide a realistic estimate; the child is unable to count the objects without touching them. The child may mark to represent their answer. b) On track – The child can contribute to the construction of a pictogram and use their addition and subtraction skills to ask and answer questions, using practical materials where necessary. Review – The child can, with support, contribute to the pictogram but is unable to ask and answer related questions. c) On track – The child can make a pattern using all three sizes of biscuit and describe it. Review – The child can copy a pattern but cannot create a pattern of their own. The child is unable to recognise the repeating unit of a pattern.

- Add the names of the children in the group/class to the top of columns 3-12. Use multiple sheets depending on the size of your class.
- Mark each child as they complete their early level end of level assessment. Suggested coding is:

  – O. On track

  – R. Review

Assessment 1: Dinosaurs

Assessment 1	Domain									
Task 1	MNU 0-02a									
Task 2	MTH 0-16a									
Task 2	MNU 0-20a									
Task 3	MNU 0-02a									
Task 3	MNU 0-11a									
Task 3	MNU 0-20a									
Task 4a, b, c	MNU 0-01a									
Task 5a, b	MTH 0-13a									
Task 6	MNU 0-07a									
Task 7	MNU 0-11a									
Task 7	MTH 0-17a MTH 0-19a									
Task 8	MNU 0-09a									

Assessment 2: The party

Assessment 2	Domain									
Task 1	MNU 0-02a									
Task 1	MNU 0-10a									
Task 2	MNU 0-02a									
Task 2	MNU 0-10a									
Task 2	MNU 0-20a									
Task 3	MNU 0-02a									
Task 3	MNU 0-07a									
Task 3	MNU 0-20a									
Task 4	MNU 0-03a									
Task 4	MNU 0-09a									

Early level end of level assessments: record sheet

Assessment 3: Playing games

Assessment 3	Domain														
Task 1	MNU 0-01a														
Task 1	MNU 0-02a														
Task 1	MNU 0-03a														
Task 2	MNU 0-13a														
Task 3	MNU 0-03a														
Task 3	MNU 0-20a														
Task 3	MNU 0-20c														
Task 4	MNU 0-20a														
Task 5	MTH 0-17a														
Task 6	MNU 0-01a														
Task 6	MNU 0-02a														
Task 6	MNU 0-09a														
Task 7	MNU 0-01a														
Task 7	MTH 0-16a														

Early level end of level assessments: record sheet

Assessment 4: One caterpillar and two caterpillars

Assessment 4	Domain								
Task 1a	MTH 0-13a								
Task 1b	MNU 0-11a								
Task 1b	MTH 0-16a								
Task 1c	MNU 0-02a								
Task 2	MNU 0-02a								
Task 2	MNU 0-10a								
Task 2	MNU 0-20c								
Task 3	MTH 0-02a								
Task 3	MTH 0-17a								
Task 3	MTH 0-19a								

Early level end of level assessments: record sheet

Assessment 5: Using the outdoors

Assessment 5	Domain												
Task 1	MNU 0-03a												
Task 1	MNU 0-11a												
Task 2	MNU 0-01a												
Task 2	MNU 0-02a												
Task 2	MNU 0-11a												
Task 3	MNU 0-01a												
Task 3	MNU 0-02a												
Task 3	MNU 0-11a												
Task 4	MTH 0-16a												
Task 5	MNU 0-02a												
Task 5	MNU 0-03a												

Assessment 6: Goldilocks' big bake-off challenge

Assessment 6	Domain										
Task 1	MNU 0-02a										
Task 1	MNU 0-10a										
Task 2	MNU 0-03a										
Task 2	MNU 0-07a										
Task 3	MNU 0-03a										
Task 3	MNU 0-07a										
Task 4	MNU 0-02a										
Task 4	MNU 0-03a										
Task 4	MTH 0-13a										
Task 5	MNU 0-01a										
Task 5	MNU 0-02a										
Task 5	MTH 0-13a										
Task 5	MNU 0-20a										